S0-ATI-531

TOPICS IN RING THEORY

MATHEMATICS LECTURE NOTE SERIES

J. Frank Adams	LECTURES ON LIE GROUPS
E. Artin and J. Tate	CLASS FIELD THEORY
Michael Atiyah	K-THEORY
Jacob Barshay	TOPICS IN RING THEORY
Hyman Bass	ALGEBRAIC K-THEORY
Melvyn S. Berger Marion S. Berger	PERSPECTIVES IN NONLINEARITY
Armand Borel	LINEAR ALGEBRA GROUPS
Raoul Bott	LECTURES ON K (X)
Andrew Browder	INTRODUCTION TO FUNCTION ALGEBRAS
Gustave Choquet	LECTURES ON ANALYSIS I. INTEGRATION AND TOPOLOGICAL VECTOR SPACES II. REPRESENTATION THEORY III. INFINITE DIMENSIONAL MEASURES AND PROBLEM SOLUTIONS
Paul J. Cohen	SET THEORY AND THE CONTINUUM HYPOTHESIS
Eldon Dyer	COHOMOLOGY THEORIES
Robert Ellis	LECTURES ON TOPOLOGICAL DYNAMICS
Walter Feit	CHARACTERS OF FINITE GROUPS
John Fogarty	INVARIANT THEORY
William Fulton	ALGEBRAIC CURVES
Marvin J. Greenberg	LECTURES ON ALGEBRAIC TOPOLOGY
Marvin J. Greenberg	LECTURES ON FORMS IN MANY VARIABLES
Robin Hartshorne	FOUNDATIONS OF PROJECTIVE GEOMETRY
J. F. P. Hudson	PIECEWISE LINEAR TOPOLOGY

Irving Kaplansky	RINGS OF OPERATORS
K. Kapp and H. Schneider	COMPLETELY O-SIMPLE SEMIGROUPS
Joseph B. Keller	BIFURCATION THEORY AND
Stuart Antman	NONLINEAR EIGENVALUE PROBLEMS
Serge Lang	ALGEBRAIC FUNCTIONS
Serge Lang	RAPPORT SUR LA COHOMOLOGIE DES GROUPES
Ottmar Loos	SYMMETRIC SPACES I. GENERAL THEORY II. COMPACT SPACES AND CLASSIFICATIONS
I. G. Macdonald	ALGEBRAIC GEOMETRY: INTRODUCTION TO SCHEMES
George W. Mackey	INDUCED REPRESENTATIONS OF GROUPS AND QUANTUM MECHANICS
Andrew Ogg	MODULAR FORMS AND DIRICHLET SERIES
Richard Palais	FOUNDATIONS OF GLOBAL NON-LINEAR ANALYSIS
William Parry	ENTROPY AND GENERATORS IN ERGODIC THEORY
D. S. Passman	PERMUTATION GROUPS
Walter Rudin	FUNCTION THEORY IN POLYDISCS
Jean-Pierre Serre	ABELIAN l-ADIC REPRESENTATIONS AND ELLIPTIC CURVES
Jean-Pierre Serre	ALGEBRES DE LIE SEMI-SIMPLE COMPLEXES
Jean-Pierre Serre	LIE ALGEBRAS AND LIE GROUPS
Shlomo Sternberg	CELESTIAL MECHANICS PART I
Shlomo Sternberg	CELESTIAL MECHANICS PART II
Moss E. Sweedler	HOPF ALGEBRAS

A Note from the Publisher

This volume was printed directly from a typescript prepared by the author, who takes full responsibility for its content and appearance. The Publisher has not performed his usual functions of reviewing, editing, typesetting, and proofreading the material prior to publication.

The Publisher fully endorses this informal and quick method of publishing lecture notes at a moderate price, and he wishes to thank the author for preparing the material for publication.

TOPICS IN RING THEORY

JACOB BARSHAY
Northeastern University

W. A. BENJAMIN, INC.
New York 1969

TOPICS IN RING THEORY

QA
247
.B36

Copyright ©1969 by W. A. Benjamin, Inc.
All rights reserved

Standard Book Numbers 8053-0550-5 (Clothbound)
8053-0551-3 (Paperback)

Library of Congress Catalog Card Number 72-99276
Manufactured in the United States of America
12345 R 32109

The manuscript was put into production on October 1, 1969;
this volume was published on December 15, 1969

W. A. BENJAMIN, INC.
New York, New York 10016

Library
UNIVERSITY OF MIAMI

To

BARBARA

PREFACE

This book is an outgrowth of a one-quarter, first-year graduate course that I taught at Northeastern University in 1966 and 1967. The lectures were based in turn on an algebra course given by Dock Sang Rim at Brandeis University in 1961-62. The book is a self-contained, general, and modern treatment of some classical theorems of commutative and noncommutative ring theory. Principally these theorems are the primary decomposition of ideals in commutative Noetherian rings and the Artin-Wedderburn structure theory for semisimple rings. By "general" and "modern" I mean that, as much as possible, theorems are proved for modules over the rings being considered and then specialized to obtain classical statements. Furthermore the techniques employed are among those which have proved fruitful in modern ring theory, for example, localization. In some sense, localization is the unifying idea in the commutative ring theory covered here.

The book begins with material usually treated in an undergraduate modern algebra course, namely, various kinds of ideals and operations on ideals, isomorphism theorems and the Chinese Remainder Theorem (Chapter 2), and Euclidean, principal ideal, and unique factorization domains (Chapter 4). However, proofs of standard theorems on unique factorization domains are not those generally given in such courses since they rely heavily on the notion of rings of quotients developed in Chapter 3. Chapter 5, an introduction to homological notions, is devoted to modules and exact sequences including the splitting of exact sequences and characterization of free and projective modules. Noetherian rings and modules are treated in Chapter 6. Since the motivation for this study is the search for a class of rings in which every ideal is a unique product of prime ideals, we are naturally led to Dedekind domains in Chapter 7. Chapters 8 and 9 are devoted to noncommutative Artin rings, including the connection between the two chain conditions by way of the idea of Jordan-Hölder series, and the structure of semisimple rings. Thus Chapters 7 and 9 can be viewed as deeper investigations of special classes of those rings studied in Chapters 6 and 8, respectively. Each chapter concludes with a set of exercises of varying degrees of difficulty.

Since the book has been expanded from the original one-quarter course of lectures, it now appears to be the appropriate amount of material for a one-semester course. Although primarily designed for beginning graduate students, it should be accessible to undergraduates who have taken the modern algebra and linear algebra courses usually offered to sophomores or juniors. For the graduate student it should provide a convenient place to learn the ring theory often expected on qualifying examinations. For the undergraduate, particularly one who is interested in algebra, the book should offer some insight into one direction his future studies might take him.

I would like to thank Professor Rim and the various authors from whom I have borrowed ideas. Their works are included in the bibliography. I would further like to acknowledge the helpful suggestions of Mark Bridger, Burton Fein, Marvin Freedman, and Kenneth Ireland. Finally, I am grateful to Delphine Radcliffe and Cindy Feldman for typing the manuscript.

JACOB BARSHAY

Cambridge, Massachusetts
July 1969

CONTENTS

Preface		v
Chapter 1.	Preliminary Terminology and Examples	1
Chapter 2.	Ideals and Residue Rings	12
Chapter 3.	Rings of Quotients and Localization	28
Chapter 4.	Unique Factorization Domains	38
Chapter 5.	Modules and Exact Sequences	52
Chapter 6.	Noetherian Rings and Modules	67
Chapter 7.	Dedekind Domains	95
Chapter 8.	Artin Rings and Modules	105
Chapter 9.	Semisimple Rings	124
Bibliography		141
Index		143

CHAPTER 1

PRELIMINARY TERMINOLOGY AND EXAMPLES

We begin with a brief discussion of just two notions from set theory. The first is that of an equivalence relation on a set and its associated decomposition; the second is Zorn's lemma. The notation used here for set membership, set inclusion, union and intersection of sets, and so forth, is standard.

DEFINITION 1-1. A binary relation $\sim$ on a set A is called an equivalence relation if for any elements a, b, c $\in$ A

(1) $a \sim a$ ($\sim$ is reflexive);

(2) if $a \sim b$, then $b \sim a$ ($\sim$ is symmetric);

(3) if $a \sim b$ and $b \sim c$, then $a \sim c$ ($\sim$ is transitive).

DEFINITION 1-2. If A is a set, $\sim$ an equivalence relation on A, and a $\in$ A, then the equivalence class of a is equal to

$\{x \in A \mid a \sim x\}$ and is denoted by $\overline{a}$.

In particular, observe that the equivalence class of an element of A is a subset of A. To say that two equivalence classes are distinct is to say that they are not equal as sets.

THEOREM 1-1. The distinct equivalence classes of an equivalence relation $\sim$ on a set A provide a decomposition of A as a union of mutually disjoint subsets.

Proof. Since $a \sim a$, we have $a \in \overline{a}$ for any $a \in A$. Thus $A \subseteq \bigcup_{a \in A} \overline{a}$. On the other hand, each $\overline{a}$ is a subset of A so $\bigcup_{a \in A} \overline{a} \subseteq A$, whence $A = \bigcup_{a \in A} \overline{a}$. To complete the proof it suffices to show that distinct equivalence classes are mutually disjoint, that is, if $a, b \in A$ then either $\overline{a} = \overline{b}$ or $\overline{a} \cap \overline{b} = \Phi$. Suppose then that $\overline{a} \cap \overline{b} \neq \Phi$ and let $x \in \overline{a} \cap \overline{b}$. Thus $a \sim x$ and $b \sim x$. But by Definition 1-1(2), $x \sim b$ and by (3) $a \sim b$. Now if $y \in \overline{b}$, then $b \sim y$ so again by (3) $a \sim y$ whence $y \in \overline{a}$. We conclude that $\overline{b} \subseteq \overline{a}$. By a similar argument, we could show $\overline{a} \subseteq \overline{b}$. Therefore $\overline{a} = \overline{b}$.

DEFINITION 1-3. A binary relation $\leq$ on a set A is called a partial ordering if for any $a, b, c \in A$

(1) $a \leq a$;

(2) if $a \leq b$ and $b \leq c$, then $a \leq c$;

(3) if $a \leq b$ and $b \leq a$, then $a = b$.

A together with the partial ordering $\leq$ is called a <u>partially ordered set</u>.

DEFINITION 1-4. A subset B of a partially ordered set A is said to be totally ordered if for any $a, b \in B$ either $a \leq b$ or $b \leq a$. A <u>totally ordered</u> subset will also be referred to as a <u>chain</u>.

DEFINITION 1-5. An element a in a partially ordered set A is called an <u>upper bound</u> for a subset B of A if for any $b \in B$, $b \leq a$.

DEFINITION 1-6. A partially ordered set A is called <u>inductive</u> if any chain in A has an upper bound in A.

DEFINITION 1-7. An element m in a partially ordered set A is called a <u>maximal element</u> if for any $a \in A$, $m \leq a$ implies $a = m$.

ZORN'S LEMMA. Every nonempty, inductive set has a maximal element.

Some applications of Zorn's lemma are in the exercises at the end of this chapter and its usefulness becomes even more apparent in later chapters.

DEFINITION 1-8. Let $f : A \to B$ be a mapping (map, function) from a set A to a set B. Then f is said to be

(1) surjective (onto) if for any element $b \in B$ there exists an element $a \in A$ such that $f(a) = b$.

(2) injective (one-to-one) if for any elements $a_1, a_2 \in A$, $f(a_1) = f(a_2)$ implies $a_1 = a_2$. [Equivalently, $a_1 \neq a_2$ implies $f(a_1) \neq f(a_2)$.]

(3) bijective (a one-to-one correspondence) if it is both surjective and injective.

DEFINITION 1-9. A group is a nonempty set G on which is defined a binary operation $*$ satisfying the following conditions:

(1) If $a, b \in G$, then $a * b \in G$. (Closure Law);

(2) If $a, b \in G$, then $(a * b) * c = a * (b * c)$. (Associative Law);

(3) There exists an element $e \in G$ such that for any $a \in G$, $e * a = a * e = a$. e is called the identity element of G.

(4) For any $a \in G$, there exists an element $\bar{a} \in G$ such that $a * \bar{a} = \bar{a} * a = e$. $\bar{a}$ is called the inverse of a.

The identity element of a group is unique as is the inverse of a given element.

DEFINITION 1-10. A group is said to be Abelian if it satisfies the additional condition:

(5) For any $a, b \in G$, $a * b = b * a$.

DEFINITION 1-11. If $(G, *)$ and $(H, \circ)$ are groups and $f : G \to H$, then f is called a group homomorphism if for any $a, b \in G$, $f(a * b) = f(a) \circ f(b)$.

DEFINITION 1-12. A ring is a set Λ on which are defined two binary operations + and $\cdot$ satisfying the following conditions:

(1) Λ is an Abelian group under +;

(2) if $a, b \in \Lambda$, then $a \cdot b \in \Lambda$ (Closure Law);

(3) if $a, b, c \in \Lambda$, then $(a \cdot b) \cdot c = a \cdot (b \cdot c)$ (Associative Law);

(4) if $a, b, c \in \Lambda$, then $a \cdot (b + c) = a \cdot b + a \cdot c$ and $(a + b) \cdot c = a \cdot c + b \cdot c$. (Distributive Laws).

There are other properties that a ring may or may not possess, among which are the following:

(5) there exists an element $1 \in \Lambda$ such that for any element $a \in \Lambda$, $1 \cdot a = a \cdot 1 = a$. 1 is called the unit element of Λ;

(6) for any element $0 \neq a \in \Lambda$, there exists an element $a^{-1} \in \Lambda$ such that $a \cdot a^{-1} = a^{-1} \cdot a = 1$. a^{-1} is called the multiplicative inverse of a;

(7) for any $a, b \in \Lambda$, $a \cdot b = b \cdot a$.

In a ring, the identity element for the operation + is denoted by 0 and the inverse of a is denoted by $-a$. The multiplication symbol $\cdot$ is generally omitted.

DEFINITION 1-13. A ring Λ satisfying

(a) (7) is called a commutative ring;

(b) (5) is called a ring with unit;

(c) (5) and (6) is called a division ring;

(d) (5) and (7) is called a commutative ring with unit;

(e) (5), (6), and (7) is called a field.

DEFINITION 1-14. If $(\Lambda, +, \cdot)$ and $(\Lambda', *, \circ)$ are rings and $f : \Lambda \to \Lambda'$, then f is called a ring homomorphism if for any $a, b \in \Lambda$, $f(a + b) = f(a) * f(b)$ and $f(a \cdot b) = f(a) \circ f(b)$.

DEFINITION 1-15. If Λ and Λ' have units 1 and 1' and $f : \Lambda \to \Lambda'$, then f is said to be unitary if $f(1) = 1'$.

DEFINITION 1-16. A group or ring homomorphism is called an

(1) epimorphism if it is surjective;

(2) monomorphism if it is injective;

(3) isomorphism if it is bijective.

EXAMPLES. 1. $\mathbb{Z} = \{0, \pm 1, \pm 2, \ldots\}$, the set of integers with + and · having the usual meaning is a commutative ring with unit element.

2. $\mathbb{Q}$, the set of rational numbers, $\mathbb{R}$, the set of real numbers, and $\mathbb{C}$, the set of complex numbers, under the usual rules of addition and multiplication are all examples of fields.

3. Let k be any field. Then k[X], the set of polynomials in one variable with coefficients in k, under the usual rules for addition and multiplication of polynomials forms a commutative ring with unit. Similarly for $k[X_1, X_2, \ldots, X_n]$, the set of polynomials in n variables with coefficients in k.

4. $\mathbb{Z}_m$, the set of integers modulo m where + and · mean addition and multiplication modulo m, forms a

commutative ring with unit element. Furthermore, $\mathbb{Z}_m$ is a field if and only if m is a prime number.

5. $M_n(k)$, the set of all n x n matrices with entries in a field k, under the usual rules for addition and multiplication of matrices, forms a ring with unit element, which is not commutative if $n \geq 2$.

6. $2\mathbb{Z} = \{0, \pm 2, \pm 4, \ldots\}$, the set of even integers, forms a commutative ring but has no unit element.

7. Δ, the real quaternions.

$$\Delta = \{x = x_0 + x_1 i + x_2 j + x_3 k \mid x_0, x_1, x_2, x_3 \in \mathbb{R}\}$$

If $x = x_0 + x_1 i + x_2 j + x_3 k$ and $y = y_0 + y_1 i + y_2 j + y_3 k$ are in Δ, then $x + y = (x_0 + y_0) + (x_1 + y_1)i + (x_2 + y_2)j + (x_3 + y_3)k$. The product xy is found by using the distributive laws and the rules $ii = jj = kk = -1$, $ij = -ji = k$, $jk = -kj = i$, and $ki = -ik = j$. Then Δ forms a division ring under these operations. In particular, the multiplicative inverse of $x = x_0 + x_1 i + x_2 j + x_3 k$ is

$$x^{-1} = \frac{x_0}{|x|} - \frac{x_1}{|x|} i - \frac{x_2}{|x|} j - \frac{x_3}{|x|} k$$

where $|x| = x_0^2 + x_1^2 + x_2^2 + x_3^2$.

EXERCISES

1-1. Show that each of the following is an equivalence relation.

(a) In the set of integers, $m \sim n$ if and only if $m - n$ is even.

(b) In the set of polynomials with real coefficients, $f(X) \sim g(X)$ if and only if α, a fixed real number, is a root of $f(X) - g(X)$.

1-2. Prove that for any two sets A and B, either there exists an injection from A to B or an injection from B to A. (Hint: Consider the set $\mathcal{H}$ of triples (X, Y, f) where $X \subseteq A$, $Y \subseteq B$, $f : X \to Y$ is a bijection. Partially order $\mathcal{H}$ by $(X_1, Y_1, f_1) \leq (X_2, Y_2, f_2)$ if and only if $X_1 \subseteq X_2$, $Y_1 \subseteq Y_2$, f_2 restricted to X_1 equals f_1. Apply Zorn's lemma and show that a maximal element of $\mathcal{H}$ must either have A as its first entry or B as its second entry.)

1-3. Let V be a vector space over a field k. Recall that a subset X of V is called linearly independent if for any finite sum $\Sigma a_i x_i = 0$ with $a_i \in k$, $x_i \in X$, all a_i must be zero. Use Zorn's lemma to prove that there exists a maximal linearly independent subset of V. Then prove that if X is such a subset and $v \in V$, then $v = \Sigma a_i x_i$ (finite sum) for some unique $a_i \in k - \{0\}$, $x_i \in X$.

1-4. $\mathbb{Z}[i] = \{a + bi \mid a, b \in \mathbb{Z}\}$. Define binary operations in the set $\mathbb{Z}[i]$ by $(a + bi) + (c + di) = (a + c) + (b + d)i$ and $(a + bi) \cdot (c + di) = (ac - bd) + (ad + bc)i$. Thus $i^2 = -1$. Prove that $\mathbb{Z}[i]$ is a commutative ring with unit. $\mathbb{Z}[i]$ is called the ring of <u>Gaussian integers</u>.

1-5. Let Λ be a ring. Prove that for each element $\lambda \in \Lambda$, the set $C(\lambda) = \{\mu \in \Lambda \mid \lambda\mu = \mu\lambda\}$ is a subring of Λ. Also prove that $C = \{\lambda \in \Lambda \mid \lambda\mu = \mu\lambda \text{ for all } \mu \in \Lambda\}$ is a commutative subring of Λ. C is called the <u>center</u> of Λ.

1-6. Let Λ be a ring and let Γ denote the set $\mathbb{Z} \times \Lambda$. Define operations in Γ by $(m, x) + (n, y) = (m + n, x + y)$ and $(m, x) \cdot (n, y) = (mn, my + nx + xy)$. Note that my should be interpreted as $y + y + \cdots + y$ (m times) in Λ. Similarly for nx. Show that Γ is a ring with unit element $(1, 0)$. Furthermore, Γ is commutative if and only if Λ is commutative. Finally consider the map $\varphi: \Lambda \to \Gamma$ given by $\varphi(x) = (0, x)$. Prove that φ is a monomorphism and that if Λ possesses a unit element, φ is not unitary.

1-7. Let $f : \Lambda \to \Lambda'$ be a homomorphism of rings with unit.

Suppose that $1' = f(\lambda)$ for some $\lambda \in \Lambda$. Prove that f is unitary.

1-8. Prove that the map $\sigma : \mathbb{Z} \to \mathbb{Z}_m$ which sends each integer to its remainder upon division by m is a ring epimorphism.

1-9. Suppose that m and n are relatively prime integers. Prove that the only ring homomorphism from $\mathbb{Z}_m$ to $\mathbb{Z}_n$ is the zero map.

1-10. Suppose that $f : \mathbb{Q} \to \mathbb{Q}$ is a ring homomorphism, not identically zero. Prove that f is the identity map.

CHAPTER 2

IDEALS AND RESIDUE RINGS

For the remainder of the book, "ring" will be understood to mean "ring with unit element." All ring homomorphisms will be assumed to be unitary.

DEFINITION 2-1. A <u>left ideal</u> A in a ring Λ is a nonempty subset of Λ such that

(1) a, b $\in$ A, then a - b $\in$ A;

(2) if a $\in$ A and $\lambda \in \Lambda$, then λa $\in$ A.

A <u>right ideal</u> A in Λ is defined by replacing condition (2) with

(2a) if a $\in$ A and $\lambda \in \Lambda$, then aλ $\in$ A.

If A satisfies (1), (2), and (2a), it is called a <u>two-sided ideal</u> or simply an <u>ideal</u>. Note that in a commutative ring (2) is equivalent to (2a) and so all ideals are two-sided.

EXAMPLES. 1. In a ring R, 0 and R are ideals. An

ideal $A \neq R$ is called <u>proper</u>.

2. In the ring of integers $\mathbb{Z}$, all multiples of a given integer n form an ideal.

3. In the ring of polynomials in one variable with real coefficients $\mathbb{R}[X]$, all polynomials that have a given real number α as a root form an ideal.

4. In a field k, the only ideals are 0 and k. For if $A \neq 0$ is an ideal of k and $a \in A$, $a \neq 0$, then $a^{-1} \cdot a = 1 \in A$ whence if $c \in k$, $c \cdot 1 = c \in A$. That is, $A = k$.

<u>Addition of ideals.</u> If A and B are left ideals in Λ then $A + B = \{a + b \mid a \in A,\ b \in B\}$ is again a left ideal of Λ called the sum of A and B.

<u>Multiplication of ideals.</u> If A and B are left ideals in Λ, then $AB = \{\Sigma_{\text{finite}}\, a_i b_i \mid a_i \in A,\ b_i \in B\}$ is again a left ideal in Λ called the product of A and B.

<u>Intersection of ideals.</u> If A_i ($i \in I$, finite or infinite) is a collection of left ideals in Λ, then $\cap_{i \in I} A_i$ is again a left ideal in Λ called the intersection of the A_i.

<u>Quotient of ideals.</u> If A and B are left ideals in Λ, then $(A{:}B) = \{\lambda \in \Lambda \mid \lambda b \in A \text{ for all } b \in B\}$ is again a left ideal in

Λ called the quotient of A by B.

At the end of the chapter, there are exercises exhibiting certain relationships among these operations.

DEFINITION 2-2. A left ideal A in a ring Λ is said to be finitely generated if there exist elements $a_1, a_2, \ldots a_n \in A$ such that every element of A can be written as $\Sigma_{i=1}^{n} \lambda_i a_i$ for some $\lambda_i \in \Lambda$. We then write $A = (a_1, a_2, \ldots, a_n)$ and call $a_1, a_2, \ldots, a_n$ a set of generators (basis, base) for A.

On the other hand, given any subset B of Λ, the set of elements that can be written as $\Sigma_{\text{finite}} \lambda_i b_i$ where $\lambda_i \in \Lambda$, $b_i \in B$ forms an ideal in Λ, denoted by (B). It is in fact the smallest ideal of Λ that contains the set B.

DEFINITION 2-3. In a commutative ring R and ideal $A = (a) = Ra$ generated by a single element is called a principal ideal. A commutative ring R in which every ideal is principal is called a principal ideal ring.

EXAMPLES. $\mathbb{Z}$ and $k[X]$ where k is a field are each principal ideal rings.

THEOREM 2-1. Let P be a proper ideal of a commutative ring R. The following conditions are equivalent:

(1) If a, b $\in$ R and ab $\in$ P, then a $\in$ P or b $\in$ P.

(2) If A and B are ideals of R and AB $\subseteq$ P, then A $\subseteq$ P or B $\subseteq$ P.

Proof. (1) implies (2). Suppose AB $\subseteq$ P but A $\not\subseteq$ P and B $\not\subseteq$ P. Then there are elements a $\in$ A, a $\notin$ P and b $\in$ B, B $\notin$ P. By (1), ab $\notin$ P. However, ab $\in$ AB $\subseteq$ P. Contradiction.

(2) implies (1). If ab $\in$ P, then (a)(b) $\subseteq$ P. Thus by (2), either (a) $\subseteq$ P or (b) $\subseteq$ P. In particular either a $\in$ P or b $\in$ P.

DEFINITION 2-4. An ideal P satisfying either (hence both) of the above conditions is called a prime ideal.

COROLLARY. Let P be a prime ideal of R. If $a_1 a_2 \cdots a_n \in$ P then some $a_i \in$ P. If $A_1 A_2 \cdots A_n \subseteq$ P, then some $A_i \subseteq$ P.

Proof. Induction on n.

THEOREM 2-2. Let $\mathfrak{m}$ be a proper left ideal of a ring A. The following conditions are equivalent:

(1) If A is a left ideal such that $\mathfrak{m} \subseteq A \subseteq \Lambda$, then $A = \mathfrak{m}$ or $A = \Lambda$.

(2) If $a \in \Lambda$, $a \notin \mathfrak{m}$, then $(\mathfrak{m}, a) = \Lambda$.

Proof. (1) implies (2). Since $a \notin \mathfrak{m}$, $A = (\mathfrak{m}, a) \supset \mathfrak{m}$. Thus $A = \Lambda$.

(2) implies (1). Suppose $\mathfrak{m} \subseteq A \subseteq \Lambda$. If $A \neq \mathfrak{m}$ then there exists $a \in A$, $a \notin \mathfrak{m}$. Thus $(\mathfrak{m}, a) = \Lambda$. But $(\mathfrak{m}, a) \subseteq A$ so $A = \Lambda$.

DEFINITION 2-5. A left ideal $\mathfrak{m}$ satisfying either (hence both) of the above conditions is called a <u>maximal left idea</u>.

THEOREM 2-3. In a commutative ring R, every maxima ideal is prime.

Proof. Suppose $\mathfrak{m}$ is a maximal ideal and $ab \in \mathfrak{m}$. If $a \notin \mathfrak{m}$, then $(\mathfrak{m}, a) = R$. In particular $1 = ra + m$ for some $r \in R$, $m \in \mathfrak{m}$. Then $b = rab + mb \in \mathfrak{m}$.

DEFINITION 2-6. If $f : \Lambda \to \Gamma$ is a ring homomorphism, then the <u>image of f</u>, denoted im f, is equal to $\{\gamma \in \Gamma \mid \gamma = f(\lambda) \text{ for some } \lambda \in \Lambda\}$; the <u>kernel of f</u>, denoted ker f, is equal to $\{\lambda \in \Lambda \mid f(\lambda) = 0\}$.

THEOREM 2-4. Let $f : \Lambda \to \Gamma$ with kernel K. Then K is and ideal of Λ.

Proof. Suppose $a, b \in K$. Then $f(a - b) = f(a) - f(b) = 0$ so $a - b \in K$. Also $f(\lambda a) = f(\lambda)f(a) = f(\lambda)0 = 0$ so $\lambda a \in K$ for any $\lambda \in \Lambda$. Similarly $f(a\lambda) = 0$ so $a\lambda \in K$. Thus K is a two-sided ideal of Λ.

Conversely, any (two-sided) ideal A of Λ is the kernel of homomorphism with domain Λ. To see this we define a relation on Λ by $a \equiv b \bmod A$, read "a congruent to b modulo A" if and only if $a - b \in A$.

THEOREM 2-5. $\equiv \bmod A$ is an equivalence relation on Λ.

Proof. The proof is immediate from the definitions of an equivalence relation and an ideal. It is thus left as an exercise.

Let Λ/A be the set of distinct equivalence classes. If $X, Y \in \Lambda/A$, say $X = \bar{a}$, $Y = \bar{b}$, then we define $X + Y = Z$ where $Z = \overline{a + b}$ and $XY = W$ where $W = \overline{ab}$.

THEOREM 2-6. Under the operations defined above, Λ/A is a ring.

Proof. It must be checked that the operations in Λ/A are well defined. In particular, if $\bar{a} = \bar{a}'$ and $\bar{b} = \bar{b}'$, we must show that $\overline{a + b} = \overline{a' + b'}$ and $\overline{ab} = \overline{a'b'}$. But $a - a' = x$ and $b - b' = y$ for some $x, y \in A$. Thus $(a + b) - (a' + b')$ $= x + y \in A$ so $\overline{a + b} = \overline{a' + b'}$. Also $ab - a'b + a'b - a'b'$ $= ab - a'b' = xb + a'y \in A$ so $\overline{ab} = \overline{a'b'}$. Hence the operations are well-defined.

Checking that the ring axioms are satisfied is left as an exercise.

There is a natural epimorphism $\sigma : \Lambda \to \Lambda/A$ given by $\sigma(\lambda) = \bar{\lambda}$. Λ/A is called the residue ring of Λ with respect to A.

THEOREM 2-7. (First Isomorphism Theorem) Suppose $f : \Lambda \to \Gamma$ is a ring homomorphism. Then $\text{im } f = \Lambda/\ker f$.

Proof. Consider the following diagram:

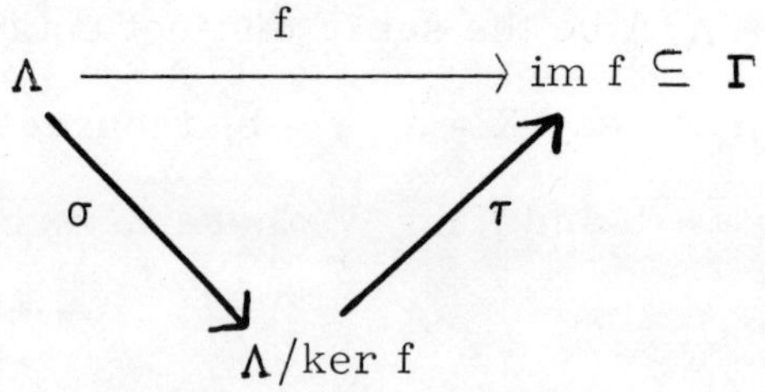

We define a map $\tau : \Lambda/\ker f \to \operatorname{im} f$ and show that it is an isomorphism. If $X = \overline{a} \in \Lambda/\ker f$, define $\tau(X) = f(a)$. To see that this is well-defined, suppose $\overline{a} = \overline{a}'$. Then $a - a' \in \ker f$ so $f(a - a') = f(a) - f(a') = 0$ or $f(a) = f(a')$. Hence $\tau(\overline{a}) = \tau(\overline{a}')$. Furthermore, if $Y = \overline{b}$, $\tau(X+Y)$ $= \tau(\overline{a} + \overline{b}) = \tau(\overline{a + b}) = f(a + b) = f(a) + f(b) = \tau(\overline{a}) + \tau(\overline{b})$ $= \tau(X) + \tau(Y)$ and $\tau(XY) = \tau(\overline{a}\overline{b}) = \tau(\overline{ab}) = f(ab) = f(a)f(b)$ $= \tau(\overline{a})\tau(\overline{b}) = \tau(X)\tau(Y)$ so τ is a homomorphism. If $\gamma \in \operatorname{im} f$, then $\gamma = f(a) = \tau(\overline{a})$ for some $a \in \Lambda$ so τ is surjective. Finally, if $\tau(\overline{a}) = f(a) = 0$, then $a \in \ker f$ so $\overline{a} = 0$. Hence τ is injective and so an isomorphism.

THEOREM 2-8. (Second Isomorphism Theorem) If $f : \Lambda \to \Lambda'$ is an epimorphism with kernel K, then there is a bijection between the set of ideals $A \supseteq K$ of Λ and the set of ideals of Λ'. Furthermore, if A and A' are corresponding ideals under this bijection, then $\Lambda/A \approx \Lambda'/A' \approx \Lambda/K/A/K$.

<u>Proof</u>. Let $\mathfrak{S}$ equal the set of ideals of Λ which contain K and $\mathfrak{J}$ equal the set of ideals of Λ'. Define $g : \mathfrak{S} \to \mathfrak{J}$ by $g(A) = \{f(a) \mid a \in A\}$, which is clearly an ideal of Λ', hence in $\mathfrak{J}$. Define $h : \mathfrak{J} \to \mathfrak{S}$ by $h(A') = \{a \in R \mid f(a) \in A'\}$ which is an ideal of Λ containing K, hence in $\mathfrak{S}$. It is easy to check that $g \circ h = I_{\mathfrak{J}}$ and $h \circ g = I_{\mathfrak{S}}$, the respective

identity maps on the sets $\mathfrak{J}$ and $\mathfrak{S}$. Hence each is a bijection.

To verify the second assertion of the theorem, let $\sigma : \Lambda' \to \Lambda'/A'$ be the natural epimorphism. Then $\tau = \sigma \circ f : \Lambda \to \Lambda'/A'$ is an epimorphism and $\lambda \in \ker \tau$ if and only if $\tau(\lambda) = 0$ if and only if $\sigma(f(\lambda)) = 0$ if and only if $f(\lambda) \in A'$ if and only if $\lambda \in A$. Hence $\ker \tau = A$ and by the previous theorem $\Lambda/A \approx \Lambda'/A'$.

DEFINITION 2-7. An element a in a commutative ring R is called a <u>zero divisor</u> if there exists $b \neq 0$ in R such that $ab = 0$. If $a \neq 0$, it is called a <u>nontrivial zero divisor</u>.

DEFINITION 2-8. A commutative ring is called an <u>integral domain</u> (or simply a domain) if it has no nontrivial zero divisors.

EXAMPLES. $\mathbb{Z}$, k, and $k[X_1, \ldots, X_n]$ where k is any field are all integral domains. On the other hand $M_n(k)$, $n \geq 2$, and $\mathbb{Z}_m$ where m is a nonprime are not integral domains.

THEOREM 2-9. Let A be an ideal of a commutative ring R.
(1) A is prime if and only if R/A is an integral domain.

(2) A is maximal if and only if R/A is a field.

<u>Proof.</u> (1) Suppose A is prime and $\bar{r}\bar{s} = 0$. Then $\overline{rs} = 0$ or $rs \in A$, whence either $r \in A$ or $s \in A$, that is, $\bar{r} = 0$ or $\bar{s} = 0$. If R/A is an integral domain and $rs \in A$, then $\bar{r}\bar{s} = 0$ so $\bar{r} = 0$ or $\bar{s} = 0$, that is, $r \in A$ or $s \in A$, whence A is prime.

(2) Suppose A is maximal and $\bar{r} \neq \bar{0}$; that is, $r \notin A$. Then $(r, A) = R$; in particular $1 = sr + a$ for some $s \in R$, $a \in A$. Then $\bar{1} = \overline{sr + a} = \overline{sr}$ so that $\bar{s}$ is the inverse of $\bar{r}$. Suppose R/A is a field and $r \notin A$. Then $\bar{r}$ has an inverse $\bar{s}$; that is, $\bar{r}\bar{s} = \bar{1}$. Thus $rs - 1 = a$ for some $a \in A$. Therefore $1 \in (r, A)$; that is, $(r, A) = R$ so A is maximal.

COROLLARY. In a commutative ring, every maximal ideal is prime.

THEOREM 2-10. A finite integral domain is a field.

<u>Proof.</u> Let $R = \{x_1, \ldots, x_n\}$ be the finite integral domain. In fact we need not assume that R has a unit element, only that it is commutative without zero divisors. Suppose $a \neq 0$ in R. Then $\mu_a : R \to R$ given by $\mu_a(x_i) = ax_i$ is a bijection. For if $i \neq j$, $x_i \neq x_j$, whence $a(x_i - x_j) \neq 0$ or $ax_i \neq ax_j$, that is, $\mu_a(x_i) \neq \mu_a(x_j)$. Since μ_a is injective and

R is finite, μ_a is bijective. Thus for some k, $1 \leq k \leq n$, $a = ax_k$. Claim: x_k is a unit element for R. For if $y \in R$, $y = ax_i$ for some i. Thus $yx_k = ax_i x_k = ax_k x_i = y$. Thus $x_k = 1$. Now $1 = ax_j$ for some value of j so a has an inverse. Therefore R is a field.

COROLLARY. If p is a prime number, $\mathbb{Z}_p$ is a field.

Proof. It suffices to show $\mathbb{Z}_p$ is an integral domain. If $\bar{a}\bar{b} = \bar{0}$, then p divides ab whence p divides a or p divides b. That is, $\bar{a} = \bar{0}$ or $\bar{b} = \bar{0}$.

Note. It is also true that a finite division ring is a field (Wedderburn's Theorem) but we will not prove this here.

DEFINITION 2-9. Let Λ_i, $i \in I$ (finite or infinite), be rings. Then the direct sum, denoted $\bigoplus_{i \in I} \Lambda_i$ is $\{(\ldots, \lambda_i, \ldots) \mid \lambda_i \in \Lambda_i$ provided that $\lambda_i = 0_i$ for all but a finite number of $i \in I\}$. Elements of $\bigoplus_{i \in I} \Lambda_i$ are added and multiplied coordinatewise making $\bigoplus_{i \in I} \Lambda_i$ into a ring (with unit if I is finite, without unit if I is infinite). Note that if I is finite, the direct sum is the same as the direct (Cartesian) product.

DEFINITION 2-10. A set of (two-sided) ideals $A_1, \ldots, A_n$ in a ring Λ is called comaximal if $A_i + A_j = \Lambda$ for $i \neq j$.

LEMMA. If A is comaximal with $A_1, \ldots, A_n$ (that is, $A + A_i = \Lambda$ for $i = 1, \ldots, n$), then A is comaximal with $\cap_{i=1}^{n} A_i$.

Proof. $\Lambda = \Lambda^n = (A + A_1)(A + A_2) \cdots (A + A_n)$

$$\subseteq A + A_1 A_2 \cdots A_n \subseteq A + \cap_{i=1}^{n} A_i \subseteq \Lambda.$$

Thus $\Lambda = A + \cap_{i=1}^{n} A_i$.

THEOREM 2-11. (Chinese Remainder Theorem) Let $A_1, \ldots, A_n$ be two-sided ideals in Λ and

$$\varphi : \Lambda \to \bigoplus_{i=1}^{n} \Lambda/A_i = \Lambda/A_1 \oplus \Lambda/A_2 \oplus \cdots \oplus \Lambda/A_n$$

be given by $\varphi(\lambda) = (\varphi_1(\lambda), \varphi_2(\lambda), \ldots, \varphi_n(\lambda))$ where $\varphi_i : \Lambda \to \Lambda/A_i$ is the natural epimorphism. Then

(1) φ is surjective if and only if $A_1, \ldots, A_n$ are comaximal;

(2) φ is injective if and only if $\cap_{i=1}^{n} A_i = 0$;

(3) φ is bijective if and only if $\cap_{i=1}^{n} A_i = 0$ and $A_1, \ldots, A_n$ are comaximal.

Proof. (1) Suppose φ is surjective and $1 \leq i \neq j \leq n$. We must show $A_i + A_j = \Lambda$. Consider $\varphi_{ij} : \Lambda \to \Lambda/A_i \oplus \Lambda/A_j$ which is surjective. Hence we can find $x_i \in A_i$ and $x_j \in A_j$ such that $\varphi_{ij}(x_i) = (\bar{0}, \bar{1})$ and $\varphi_{ij}(x_j) = (\bar{1}, \bar{0})$. Then $\varphi_{ij}(1 - (x_i + x_j)) = (\bar{1}, \bar{1}) - (\bar{0}, \bar{1}) - (\bar{1}, \bar{0}) = (\bar{0}, \bar{0})$ so $1 - (x_i + x_j) \in \ker \varphi_{ij} = A_i \cap A_j \subseteq A_i + A_j$. But $x_i + x_j \in A_i + A_j$ as well so $1 \in A_i + A_j$. That is, $\Lambda = A_i + A_j$.

Now suppose $A_1, \ldots, A_n$ is a comaximal set. We must show φ is surjective, that is, given $\lambda_1, \ldots, \lambda_n \in \Lambda$, there exists an element $x \in \Lambda$ such that $\varphi_i(x) = \bar{\lambda}_i$ for $i = 1, \ldots, n$. This is clear for $n = 1$. Suppose $n = 2$. Since $A_1 + A_2 = \Lambda$, there exists $e_1 \in A_1$, $e_2 \in A_2$ such that $1 = e_1 + e_2$. Set $x = e_1\lambda_2 + e_2\lambda_1$. Then $\varphi_1(x) = \bar{e}_1\bar{\lambda}_2 + \bar{e}_2\bar{\lambda}_1 = \bar{e}_2\bar{\lambda}_1 = \overline{(1 - e_1)}\bar{\lambda}_1 = \bar{\lambda}_1 - \bar{e}_1\bar{\lambda}_1 = \bar{\lambda}_1$ and similarly, $\varphi_2(x) = \bar{\lambda}_2$.

Assume now that we have established the result for $n = 1, 2, \ldots, k-1$. By the lemma A_k is comaximal with $\cap_{i=1}^{k-1} A_i$. Thus we can find $y \in \cap_{i=1}^{k-1} A_i$ and $z \in A_k$ such that $1 = y + z$. By the induction hypothesis, there exists an element $a \in \Lambda$ such that $\varphi_i(a) = \bar{\lambda}_i$, $i = 1, \ldots, k-1$. Set $x = za + \lambda_k y$. Then $\varphi_k(x) = \bar{\lambda}_k$ since $z \in A_k$ and $y = 1 - z$. Also $\varphi_i(x) = \varphi_i(a) = \bar{\lambda}_i$ for $i = 1, \ldots, k-1$ since $y \in \cap_{i=1}^{k-1} A_i \subseteq A_i$ and $z = 1 - y$. Thus x is the desired element, completing the induction. (2) and (3) are trivial since $\ker \varphi = \cap_{i=1}^{n} A_i$.

COROLLARY. A system of linear congruences

$$X \equiv c_i \pmod{m_i} \qquad i = 1, \ldots, t$$

is solvable for all values of $c_1, \ldots, c_t$ if and only if the moduli are relatively prime in pairs.

Proof. To state that the system is solvable for all values of $c_1, \ldots, c_t$ is to state that the natural map from $\mathbb{Z} \to \mathbb{Z}_{m_1} \oplus \mathbb{Z}_{m_2} \oplus \cdots \oplus \mathbb{Z}_{m_t}$ is surjective. By the previous theorem, that means the ideals $(m_1), (m_2), \ldots, (m_t)$ must be comaximal. But $1 \in (m_i) + (m_j)$ if and only if $(m_i, m_j) = 1$, that is, the moduli must be relatively prime in pairs.

EXERCISES

2-1. Let A, B, C be left ideals in a ring Λ. Prove each of the following:

(a) $A(B+C) = AB + AC$

(b) $(A:B)B \subseteq A$

(c) $(A:(B+C)) = (A:B) \cap (A:C)$

(d) $((A:C):B) = (A:BC)$

2-2. Let A, B, C be two-sided ideals in a ring Λ. Prove each of the following:

(a) $AB \subseteq A \cap B$

(b) $A \subseteq (A:B)$

2-3. Let Λ be the ring of continuous real-valued functions on $[0, 1]$ where the operations in Λ are given by $(f+g)(t) = f(t) + g(t)$ and $(fg)(t) = f(t)g(t)$. For any t, $0 \leq t \leq 1$, define $\tilde{t} : \Lambda \to \mathbb{R}$ by $\tilde{t}(f) = f(t)$. Prove that $\ker \tilde{t}$ is a maximal two-sided ideal of Λ.

2-4. Complete the proof of Theorem 2-8. That is, show that $g \circ h$ and $h \circ g$ are the appropriate identity maps.

2-5. Let A be the principal ideal generated by $X^2 + 1$ in the ring $R = \mathbb{Z}[X]$. Prove that $R/A \approx \mathbb{Z}[i]$, the Gaussian integers.

2-6. Show by example that $M_n(\mathbb{R})$ is not an integral domain for $n \geq 2$.

2-7. Prove that in a finite commutative ring, every prime ideal is maximal.

2-8. Suppose that A and B are comaximal ideals in a

commutative ring R. Prove that $AB = A \cap B$. Generalize to sets of n comaximal ideals.

2-9. Let $A_1, \ldots, A_n$ be comaximal two-sided ideals in Λ. Prove that $A_1^{m_1}, \ldots, A_n^{m_n}$ are comaximal for any positive integers $m_1, \ldots, m_n$.

2-10. Let $n = p_1^{m_1} p_2^{m_2} \cdots p_k^{m_k}$ be the prime factorization of the integer n. Prove that

$$\mathbb{Z}_n \approx \mathbb{Z}_{p_1^{m_1}} \oplus \cdots \oplus \mathbb{Z}_{p_k^{m_k}}.$$

CHAPTER 3

RINGS OF QUOTIENTS AND LOCALIZATION

In this chapter we discuss another construction yielding a new ring from a given ring. The reader should keep in mind the method by which the rationals $\mathbb{Q}$ are constructed from the integers $\mathbb{Z}$ for it is just this process which is being generalized. In this chapter and the next, all rings are assumed commutative.

DEFINITION 3-1. A subset S of a ring R is called a <u>multiplicative set</u> if

(1) $1 \in S$;

(2) if $a, b, \in S$, then $ab \in S$.

Let S be a multiplicative set in R. Consider the set $\{r/s \mid r \in R, s \in S\}$ thought of simply as formal symbols. We say two such symbols r_1/s_1 and r_2/s_2 are equivalent, denoted $r_1/s_1 \sim r_2/s_2$, if there exists $s \in S$ such that $s(r_1 s_2 - r_2 s_1) = 0$. The reader should check that $\sim$ is in fact an equivalence relation. Denote by $[r/s]$ the class of r/s and by R_S the set of distinct equivalence classes. We define addition and multiplication in R_S by

$$[r_1/s_1]+[r_2/s_2] = [r_1 s_2 + r_2 s_1/s_1 s_2]$$

and

$$[r_1/s_1] \cdot [r_2/s_2] = [r_1 r_2/s_1 s_2] .$$

Under these operations R_S is a ring called the ring of quotients of R with respect to S. Furthermore there is a natural homomorphism $\varphi : R \to R_S$ given by $\varphi(r) = [r/1]$.

THEOREM 3-1. (1) If $0 \in S$, then $R_S = 0$.
(2) φ is injective if and only if S contains no zero divisors.

Proof. (1) Note that $[0/1]$ is the zero element of R_S since $[r/s] + [0/1] = [r1 + 0s/s1] = [r/s]$. If $0 \in S$ and $[r/s] \in R_S$, then $[r/s] = [0/1]$ since $0(r1 + 0s) = 0$. Thus R_S reduces to just the zero element.

(2) $r \in \ker \varphi$ if and only if $\varphi(r) = [r/1] = 0$ in R_S if and only if there exists $s \in S$ such that $sr = 0$. Thus $\ker \varphi = 0$ if and only if S contains no zero divisors.

EXAMPLES. 1. Let S be the set of all non-zero divisors of R. Suppose a, b $\in$ S and c $\in$ R satisfy (ab)c = 0. Then a(bc) = 0 which implies bc = 0 since a $\in$ S. But this implies c = 0 since b $\in$ S. Thus ab $\in$ S. Clearly 1 $\in$ S so S is a multiplicative set. R_S is called the total ring of quotients of R.

2. Let P be a prime ideal of R and S = R - P. This is a multiplicative set. The ring of quotients R_S is usually denoted by R_P and is called the localization of R at P.

3. As a special case of either example 1 or 2, let R be an integral domain. Then (0) is a prime ideal and S = R - (0) is the set of all non-zero divisors of R. The localization at (0) is called the quotient field of R. As the name suggests, it is in fact a field.

THEOREM 3-2. Let S be a multiplicative set in a ring R. Then there exists a bijection between the set of prime ideals of R whose intersection with S is empty and the set of prime ideals of R_S.

Proof. If 0 $\in$ S, then both sets are empty. Thus we can assume 0 $\notin$ S. We begin by describing a method of associating an ideal in R_S with one in R and vice versa.

If A is an ideal in R, define

$$AR_S = \{t[a/1] \mid a \in A,\ t \in R_S\}$$

This is called the extension of A to R_S. On the other hand, if B is an ideal in R_S, define

$$B \cap R = \varphi^{-1}(B)$$

where $\varphi : R \to R_S$ is the natural homomorphism. This is called the contraction of B to R. Note that when S contains zero divisors, φ is not injective so that R cannot be thought of as embedded in R_S. In this case the contraction is not a genuine intersection. However, the intersection notation is a widely accepted one.

The proof can now be broken down into a sequence of steps.

(a) AR_S is an ideal of R_S. For if $t_1[a_1/1]$ and $t_2[a_2/1]$ are in AR_S where $t_1 = [r_1/s_1]$ and $t_2 = [r_2/s_2]$, then

$$\begin{aligned} t_1[a_1/1] - t_2[a_2/1] &= [r_1 a_1/s_1] - [r_2 a_2/s_2] \\ &= [r_1 a_1 s_2 - r_2 a_2 s_1/s_1 s_2] = t'[a'/1] \end{aligned}$$

where $t' = [1/s_1 s_2] \in R_S$ and $a' = r_1 a_1 s_2 - r_2 a_2 s_1 \in A$.

(b) If P is prime in R and $P \cap S = \Phi$, then PR_S is prime in R_S. First of all PR_S is a proper ideal of R_S. For if $1 \in PR_S$, then $[1/1] = [rp/s]$ for some $r \in R$, $p \in P$,

$s \in S$ in which case there exists $s' \in S$ such that $s'(s - rp) = 0$. That is $s's = s'rp$. But $s's \in S$ and $s'rp \in P$ so $P \cap S \neq \Phi$. Contradiction. Thus PR_S is proper. Furthermore if $[r_1/s_1][r_2/s_2] = [r_1 r_2/s_1 s_2] \in PR_S$, say $[r_1 r_2/s_1 s_2] = [rp/s]$ for some $r \in R$, $s \in S$, $p \in P$, then there exists $s' \in S$ such that $s'(r_1 r_2 s - rps_1 s_2) = 0$. Thus $s'r_1 r_2 s = s'rps_1 s_2 \in P$. But $s \notin P$, $s' \notin P$ so either $r_1 \in P$ or $r_2 \in P$. Hence either $[r_1/s_1] \in PR_S$ or $[r_2/s_2] \in PR_S$.

(c) $B \cap R$ is an ideal of R. If $b_1, b_2 \in B \cap R$, then $\varphi(b_1) = [b_1/1]$ and $\varphi(b_2) = [b_2/1] \in B$. Thus $[b_1/1] - [b_2/1] = [b_1 - b_2/1] = \varphi(b_1 - b_2) \in B$. Hence $b_1 - b_2 \in B \cap R$.

(d) If B is prime in R_S, then $B \cap R$ is prime in R and $(B \cap R) \cap S = \Phi$. For $ab \in B \cap R$ implies $[ab/1] = [a/1][b/1] \in B$ whence $[a/1] \in B$ or $[b/1] \in B$, that is, $a \in B \cap R$ or $b \in B \cap R$. Furthermore, if $s \in (B \cap R) \cap S$, then $[s/1] \in B$ so $[1/s][s/1] = [1/1] \in B$ which implies $B = R_S$. Contradiction. Thus $(B \cap R) \cap S = \Phi$. In particular, $1 \notin B \cap R$ so $B \cap R$ is proper, hence prime.

(e) It remains only to show that this pairing is actually a bijection between the two sets in question. This is left as an exercise for the reader.

DEFINITION 3-2. An element r in a ring Λ (not necessarily commutative) is called a unit if there exists $s \in \Lambda$ such that $rs = 1 = sr$.

THEOREM 3-3. The following statements are equivalent:

(1) The set of nonunits of R form an ideal.

(2) R has a unique maximal ideal.

Proof. (1) implies (2). Let M be the ideal of nonunits. If $x \notin M$, then x is a unit so $R = (x) \subseteq (x, M) \subseteq R$. Thus $(x, M) = R$ which implies M is maximal. Now suppose $\mathfrak{m}$ is any maximal ideal of R. Then $\mathfrak{m}$ consists solely of nonunits. Thus $\mathfrak{m} \subseteq M \subset R$ which implies $\mathfrak{m} = M$.

(2) implies (1). We first show that any proper ideal of R is contained in at least one maximal ideal. To do this we will use Zorn's lemma. Let A be a proper ideal of R and consider

$$\mathcal{S} = \{B \mid B \text{ is an ideal of } R \text{ satisfying } A \subseteq B \subset R\}.$$

$\mathcal{S}$ is not empty since A is in $\mathcal{S}$. Furthermore, if B_i, $i \in I$, is a chain in $\mathcal{S}$, then $B = \bigcup_{i \in I} B_i$ is again in $\mathcal{S}$. For certainly $A \subseteq B$ and if $B = R$, then $1 \in B$, in which case $1 \in B_i$ for some i. Thus $B_i = R$ contradicting the assumption that B_i is in $\mathcal{S}$. B is an upper bound for the chain so

$\mathcal{S}$ is an inductive set. By Zorn's lemma, let M be a maximal element of $\mathcal{S}$. If $x \in R$, $x \notin M$, then $M \subset (x, M) \subseteq R$. By the choice of M, it must be that (x, M) is not in $\mathcal{S}$, that is, $(x, M) = R$. Thus M is a maximal ideal of R which contains A.

Now let $\mathfrak{m}$ denote the unique maximal ideal of R and M the set of nonunits of R. Clearly $\mathfrak{m} \subseteq M$ since $\mathfrak{m}$ consists solely of nonunits. On the other hand, if $x \in M$, then $(x) \subseteq \mathfrak{m}$ since $\mathfrak{m}$ is the only maximal ideal. In particular, $x \in \mathfrak{m}$ so $M \subseteq \mathfrak{m}$. Therefore $M = \mathfrak{m}$ and so M is an ideal.

DEFINITION 3-3. A ring satisfying either (hence both) of the above conditions is called a <u>local ring</u>.

THEOREM 3-4. Let P be a prime ideal of R. Then R_P is a local ring with unique maximal ideal PR_P.

<u>Proof</u>. By Theorem 3-2, the only prime ideals of R_P are of the form QR_P where Q is a prime ideal of R and $Q \cap (R - P) = \Phi$, that is, $Q \subseteq P$. Thus $QR_P \subseteq PR_P$. Since maximal ideals are prime, PR_P must be the only maximal ideal of R_P.

THEOREM 3-5. Let P be a prime ideal of R. Then R_P/PR_P is isomorphic to the quotient field of R/P.

Proof. Recall that the quotient field of an integral domain is just the localization at the prime ideal 0. Define a map $\rho : (R/P)_0 \to R_P/PR_P$ by

$$\rho([\tilde{r}/\tilde{s}]) = \overline{[r/s]}$$

where $\varphi : R \to R/P$ sends $r \to \tilde{r}$, $\psi : R_P \to R_P/PR_P$ sends $t \to \bar{t}$, and [ ] has the usual meaning of the class of an element in a ring of quotients. We must show that ρ is well-defined and an isomorphism.

(a) ρ is well-defined. Suppose $[\tilde{r}_1/\tilde{s}_1] = [\tilde{r}_2/\tilde{s}_2]$. Then there exists $\tilde{r} \neq 0$ in R/P such that $\tilde{r}(\tilde{r}_1\tilde{s}_2 - \tilde{r}_2\tilde{s}_1) = 0$. That is, there exists $r \in R - P$ such that $r(r_1 s_2 - r_2 s_1) \in P$. Thus $r_1 s_2 - r_2 s_1 \in P$ which implies $[r_1 s_2 - r_2 s_1/s_1 s_2] \in PR_P$. Therefore $\overline{[r_1 s_2 - r_2 s_1/s_1 s_2]} = 0$; that is $\overline{[r_1/s_1]} = \overline{[r_2/s_2]}$.

(b) ρ is a ring homomorphism. This verification is left as an exercise for the reader.

(c) Suppose $[\tilde{r}_1/\tilde{s}_1] \in \ker \rho$. Then $[r_1/s_1] \in PR_P$. Thus there exist elements $p \in P$, $r \in R$, $s \in R - P$ such that $[r_1/s_1] = [rp/s]$, whence there exists $s' \in R - P$ such that $s'sr_1 = s'rps_1 \in P$. But $s \notin P$, $s' \notin P$ so $r_1 \in P$. Therefore $\tilde{r}_1 = 0$ and so $[\tilde{r}_1/\tilde{s}_1] = 0$. Hence ρ is injective. It is immediate from the definition that ρ is surjective, hence an isomorphism.

EXERCISES

3-1. Prove that the complement of a union of prime ideals in a ring R is a multiplicative set.

3-2. Let k be a field, $a \in k$, and set

$$M_a = \{f(X) \in k[X] \mid f(a) \neq 0\}.$$

Show that M_a is a multiplicative set in the ring $k[X]$. More generally, let V be any collection of n-tuples $(a) = (a_1, \ldots, a_n)$ in k^n and set

$$M_V = \{f(X_1, \ldots, X_n) \in k[X_1, \ldots, X_n] \mid f(a) \neq 0 \text{ for all } (a) \in V\}.$$

Show that M_V is a multiplicative set in $k[X_1, \ldots, X_n]$.

3-3. Prove that the quotient field of $\mathbb{Z}[i]$, the Gaussian integers, is isomorphic to $\mathbb{Q}[i] = \{a + bi \mid a, b \in \mathbb{Q}\}$.

3-4. Describe the total ring of quotients of $\mathbb{Z}_n$.

3-5. Complete the proofs of Theorem 3-2(e) and

Theorem 3-5(b).

3-6. Prove that the set of units of a ring form a group under multiplication.

3-7. Find all units in the following rings:

(a) $\mathbb{Z}[i]$

(b) $k[X]$

(c) $\mathbb{Z}_n$

(d) $M_2(\mathbb{R})$

3-8. Let R be an integral domain with quotient field K, S a multiplicative set in R, $0 \notin S$. Prove that R_S is an integral domain and that the quotient field of R_S is K.

3-9. Let S be a multiplicative subset of a ring R, $0 \notin S$. Let P be a maximal element in the set of ideals whose intersection with S is empty. (Show by Zorn's lemma that there exists such an ideal.) Prove that P is a prime ideal.

3-10. Let R be a ring in which every nonzero prime ideal is maximal. Prove that PR_P is the only nonzero prime ideal of R_P where $P \neq 0$ is a prime ideal of R.

CHAPTER 4

UNIQUE FACTORIZATION DOMAINS

In this chapter we will employ the technique of localization developed in the previous chapter to capture some well known results about unique factorization domains, namely Theorems 4-6, 4-7, and 4-8. We begin the chapter with a special class of these rings called Euclidean domains. Once again, all rings are commutative.

DEFINITION 4-1. An integral domain R is called a <u>Euclidean domain</u> if there exists a function $d : R \to \mathbb{Z}$ satisfying

(1) $d(a) > d(0)$ for all $0 \neq a \in R$;

(2) For any $a, b \in R$, $b \neq 0$, there exist elements $q, r \in R$ such that $a = qb + r$ with $d(r) < d(b)$.

EXAMPLES. 1. Let $R = \mathbb{Z}$ and $d(a) = |a|$, ordinary absolute value. The elements q and r are what are usually called the quotient and remainder upon division of a by b.

2. Let R = k[X] for a field k and set d(f(X)) = the degree of the polynomial f(X) if f(X) $\neq$ 0 and d(0) = -1.

THEOREM 4-1. Every Euclidean domain is a principal ideal domain.

Proof. Let (R, d) be a Euclidean domain and B an ideal in R. If B is the zero ideal, then B = (0) and so is principal. Otherwise consider the non-empty subset X of $\mathbb{Z}$ given by X = $\{d(a) \mid a \in B, a \neq 0\}$. By property (1) of the function d, x > d(0) for all x $\in$ X. Thus X is a nonempty subset of $\mathbb{Z}$ which is bounded from below. Hence X has a minimal element. Let 0 $\neq$ b $\in$ B be such that d(b) is a minimal element of X. We want to show that B = (b).

Suppose a $\in$ B. Then there exist elements q and r in R such that a = qb + r with d(r) < d(b). Since a $\in$ B, qb $\in$ B, we have r = a - qb $\in$ B. But d(r) < d(b) contradicts the choice of b unless r = 0. Hence a = qb and B = (b). Therefore every ideal of R is principal.

DEFINITION 4-2. A ring R is said to satisfy the <u>ascending chain condition</u> if every strictly ascending chain of ideals of R, $A_1 \subset A_2 \subset A_3 \subset \cdots$ is finite. Equivalently if for every infinite chain of ideals $A_1 \subseteq A_2 \subseteq \cdots$, there exists an integer k such that $A_i = A_k$ for $i \geq k$.

DEFINITION 4-3. A ring R is said to satisfy the maximum condition if every nonempty collection of ideals of R has a maximal element, that is, an ideal which is properly contained in no ideal of the collection.

THEOREM 4-2. A ring R satisfies the ascending chain condition if and only if R satisfies the maximum condition.

Proof. If R does not satisfy the ascending chain condition, there exists an infinite strictly ascending chain of ideals $\{A_i\}$. The collection of these ideals has no maximal element.

If R does satisfy the ascending chain condition, let $\mathcal{H}$ be a nonempty collection of ideals. Let $A_1 \in \mathcal{H}$. If A_1 is maximal, we are done. Otherwise, there exists $A_2 \in \mathcal{H}$ such that $A_1 \subset A_2$. If A_2 is maximal, we are done. Otherwise, continue the process. Since R satisfies the ascending chain condition, this process must stop. When it does, we have a maximal element in $\mathcal{H}$.

To say that a ring satisfies the ascending chain condition for principal ideals has the obvious meaning, that is, replace "ideal" by "principal ideal" in Definition 4-2.

DEFINITION 4-4. Let R be a domain. A nonunit $p \in R$ is called irreducible if $p = ab$ implies either a or b is a unit in R.

DEFINITION 4-5. An integral domain R is called a unique factorization domain if every nonzero nonunit of R can be written uniquely as a finite product of irreducibles. More precisely,

(1) If $a \neq 0$ is a nonunit, then $a = p_1 p_2 \cdots p_r$ where each p_i is irreducible.

(2) If $p_1 \cdots p_r = q_1 \cdots q_s$ (all p_i and q_j irreducible) then $r = s$ and there exists a permutation π of $\{1, 2, \ldots, r\}$ such that $p_i = u_i q_{\pi(i)}$ for some units u_i.

DEFINITION 4-6. A nonzero element $p \in R$ is called a prime if (p) is a prime ideal of R.

Note. 1. Every prime is irreducible. For if p is prime and $p = ab$, then $ab \in (p)$ so either $a \in (p)$ or $b \in (p)$. If

$a \in (p)$, then $a = rp$ for some $r \in R$. Thus $p = ab = rpb$, that is, $1 = rb$ so b is a unit in R.

2. Not every irreducible is prime. Consider the ring $\mathbb{Z}[\sqrt{-5}] = \{a + b\sqrt{-5} \mid a, b \in \mathbb{Z}\}$ where the operations are the usual ones for complex numbers. In this ring, $2 \cdot 3 = (1 + \sqrt{-5})(1 - \sqrt{-5})$. All of these elements are irreducible, but none is prime.

THEOREM 4-3. A domain R is a unique factorization domain if and only if every nonzero nonunit of R can be written as a finite product of prime elements.

Proof. Assume R is a unique factorization domain. It suffices to show that every irreducible element of R is prime. Let p be irreducible and suppose $ab \in (p)$, that is, $ab = rp$ for some $r \in R$. Write a, b, r as products of irreducibles, say $a = p_1 p_2 \cdots p_\alpha$, $b = q_1 q_2 \cdots q_\beta$, $r = p_1' p_2' \cdots p_\rho'$. Then $p_1 p_2 \cdots p_\alpha q_1 \cdots q_\beta = p_1' p_2' \cdots p_\rho' p$. By uniqueness $p = up_i$ for some i, $1 \le i \le \alpha$ or $p = vq_j$ for some j, $1 \le j \le \beta$, u and v units. In one case $a \in (p)$, in the other $b \in (p)$. Therefore (p) is a prime ideal so p is a prime element.

Now assume every nonunit can be written as a finite product of primes. Since primes are irreducible, it suffices to show the expression is unique. Suppose $p_1 \cdots p_r = q_1 \cdots q_s$ where all p_i, q_j are primes. Then p_1 divides q_j for some j, say j = 1 (relabel if necessary). Hence $q_1 = u_1 p_1$ where u_1 is a unit. After cancellation $p_2 \cdots p_r = u_1 q_2 \cdots q_s$. Proceeding by induction, the uniqueness follows.

THEOREM 4-4. Let S be the multiplicative set generated by 1 and all primes in the domain R. Then R is a unique factorization domain if and only if R_S is a field.

Proof. Assume R is a unique factorization domain. Then every nonzero nonunit of R is in S. But in R_S, elements of S become units. Hence every nonzero element of R_S is a unit, that is, R_S is a field.

Suppose R is not a unique factorization domain and $a \in R$ is a nonzero nonunit which cannot be written as a finite product of primes, that is, $a \notin S$. Clearly then $(a) \cap S = \Phi$. For if ba is a finite product of primes, a must be also. Thus $0 \neq (a)R_S \neq R_S$, that is, the ideal generated by $[a/1]$ in R_S is nonzero and proper. Therefore R_S is not a field since it has a nonzero proper ideal.

THEOREM 4-5. Let R be a domain with the ascending

chain condition on principal ideals. Let M be a multiplicative set generated by 1 and prime elements (not necessarily all prime elements). If R_M is a unique factorization domain, then R is a unique factorization domain.

Proof. Let S be the multiplicative set generated by 1 and all primes in R and let T be the multiplicative set generated by $[1/1]$ and all primes in R_M. We will accomplish the proof by a series of reductions.

(a) Since R_M is by hypothesis a unique factorization domain, Theorem 4-4 states that $(R_M)_T$ is a field. Again by Theorem 4-4, R is a unique factorization domain if and only if R_S is a field. Thus it suffices to prove $R_S = (R_M)_T$.

(b) $R_S \subseteq (R_M)_T$. For let $x = [r/s] \in R_S$. Write $s = ms'$ where m is a product of generators of M and s' involves no generators of M. Then $x = [r/m][1/s']$ where clearly $[r/m] \in (R_M)_T$. Thus it suffices to show that $[1/s'] \in (R_M)_T$. Furthermore it suffices to show that $[1/p] \in (R_M)_T$ where p is a prime of R and not one of the generators of M, since $[1/s']$ is just a product of such elements. Now (p) is a prime ideal of R. If $(p) \cap M \neq \Phi$, then $rp \in M$ for some $r \in R$. Thus $[1/rp] \in R_M$ so $[1/p] = [r/1][1/rp] \in R_M \subseteq (R_M)_T$ and we are done.

Otherwise $(p) \cap M = \Phi$ in which case $[p/1]$ generates a prime ideal in R_M. That is, $[p/1]$ is a prime element in R_M, hence in T. Thus $[[1/1]/[p/1]] = [1/p] \in (R_M)_T$. This shows that $R_S \subseteq (R_M)_T$.

(c) The next claim is that in order to show $(R_M)_T \subseteq R_S$, it suffices to prove the following statement: if $x \in R$ and $[x/1]$ is a prime in R_M, then $x \in S$. For a typical element of $(R_M)_T$ can be written as $[z/t]$ where $z \in R_M$, $t \in T$ and $t = [x_1/1][x_2/1]\cdots[x_k/1]$ where $x_1, \ldots, x_k \in R$ and $[x_1/1], \ldots, [x_k/1]$ are prime in R_M. Our statement would then say that $x_1, \ldots, x_k \in S$. Thus $[1/t] \in R_S$. But $z \in R_M \subseteq R_S$ (since $M \subseteq S$) so we can conclude $[z/t] \in R_S$. This verifies the claim.

(d) If $x \in R$ and $[x/1]$ is a prime in R_M, then $x \in S$. Assume the contrary, that is, there exists an element $x \in R - S$ such that $[x/1]$ is prime in R_M. Let $\mathcal{S} = \{(x) \subset R \mid x \in R - S \text{ and } [x/1] \text{ is prime in } R_M\}$. By hypothesis $\mathcal{S}$ is nonempty and by the ascending chain condition on principal ideals, there exists a maximal element in $\mathcal{S}$. Call it (y).

The next claim is that $[y/1]R_M \cap R = (y)$. Suppose that $[y/1][r/m] \in R$ for some $r \in R$, $m \in M$. That is, m divides yr. We want to show that m divides r. If p is a prime (in R) factor of m and p divides y, then $y = pz$ for some $z \in R$. Clearly $z \notin S$ for $z \in S$ would imply $y \in S$. Furthermore $[z/1]R_M = [y/p]R_M = [y/1]R_M$ so $[z/1]$ is prime in R_M. Hence (z) is in $\mathcal{S}$ and by the choice of (y), $(z) = (y)$. Thus $z = ay$ for some a in R from which $y = pz = pay$. Therefore $pa = 1$ making p a unit and contradicting p a prime. Thus no prime factor of m divides y, whence m divides r. This gives the inclusion $[y/1]R_M \cap R \subseteq (y)$. The reverse inclusion is immediate and the claim is established.

Finally we conclude from the claim and Theorem 3-2 that y is a prime in R. This immediately contradicts $y \notin S$ and completes the proof of the theorem.

THEOREM 4-6. Every principal ideal domain is a unique factorization domain.

Proof. Let R be a principal ideal domain and S the multiplicative set generated by 1 and all primes in R. If R_S is not a field, let $0 \neq A \subset R_S$ be a maximal ideal in R_S. Then A is a prime ideal of R_S so $A \cap R$ is a prime ideal of R. But $A \cap R = (r)$ for some $r \in R$, whence r is prime in R. Thus $r \in S$. But this implies $A = (A \cap R)R_S = [r/1]R_S = R_S$, contradicting the choice of A. Thus R_S is a field and by Theorem 4-4, R is a unique factorization domain.

COROLLARY. Every Euclidean domain is a unique factorization domain.

LEMMA. (1) If R is a domain, then R[X] is a domain. (2) If R satisfies the ascending chain condition for principal ideals, so does R[X].

<u>Proof</u>. (1) Obvious. (2) Consider $(f_1(X)) \subseteq (f_2(X)) \subseteq \cdots$. Then $\deg f_1(X) \geq \deg f_2(X) \geq \cdots$. This must end at some nonnegative integer. Suppose $\deg f_i(X) = \deg f_k(X)$ for all $i \geq k$. Then $(f_k(X)) \subseteq (f_{k+1}(X))$ implies $f_k(X) = af_{k+1}(X)$ for some $a \in R$.

Let a_i be the leading coefficient of $f_i(X)$. Then $(a_k) \subseteq (a_{k+1}) \subseteq \cdots$. Thus there exists N such that $(a_j) = (a_t)$ for $j, t \geq N$. Suppose $j > t > N$. Then $f_j(X)$ divides $f_t(X)$, that is, $f_t(X) = af_j(X)$. Therefore $a_t = aa_j$. But $(a_t) = (a_j)$ so a is a unit in R. Therefore $(f_t(X)) = (f_j(X))$.

THEOREM 4-7. If R is a unique factorization domain, then R[X] is a unique factorization domain.

<u>Proof</u>. Note that for any ideal A or R, $(R/A)[X] \approx R[X]/AR[X]$. Therefore if P is a prime ideal in R, then PR[X] is a prime

ideal in $R[X]$. So if p is a prime element in R, it is also a prime element in $R[X]$. Let S be the multiplicative set generated by prime elements in R.

Then $(R[X])_S = R_S[X]$. But R_S is a field so $R_S[X]$ is a principal ideal domain. Then by Theorem 4-6, $R_S[X]$ is a unique factorization domain, whence by Theorem 4-5 and the above lemma, $R[X]$ is a unique factorization domain.

Note. We have used here the fact that a unique factorization domain satisfies the ascending chain condition on principal ideals. This follows immediately by considering the factorization of the generators of the ideals in a chain.

COROLLARY. If R is a unique factorization domain, then $R[X_1, \ldots, X_n]$ is a unique factorization domain.

COROLLARY. If k is a field, $k[X_1, \ldots, X_n]$ is a unique factorization domain.

Note. Not every unique factorization domain is a principal ideal domain. For example, $k[X_1, \ldots, X_n]$, $n \geq 2$.

DEFINITION 4-7. If $a = p_1^{\alpha_1} \cdots p_t^{\alpha_t}$ and $b = p_1^{\beta_1} \cdots p_t^{\beta_t}$ are prime factorizations of a and b in the unique factorization domain R where $\alpha_i \geq 0$, $\beta_j \geq 0$, then $d = \prod_{i=1}^{t} p_i^{\min(\alpha_i, \beta_i)}$ is called a greatest common divisor (g. c. d) of a and b. It is unique up to multiplication by a unit.

DEFINITION 4-8. Let R be a unique factorization domain and $f(X) = a_0 + a_1 X + \cdots + a_n X^n \in R[X]$. Then the content of $f = c(f) = \text{g.c.d.}(a_0, \ldots, a_n)$. If $c(f) = 1$, f is called a primitive polynomial.

THEOREM 4-8. (Gauss lemma) Let R be a unique factorization domain with quotient field K. If $f(X) \in R[X]$ is irreducible over $R[X]$, then it is irreducible over $K[X]$.

Proof. Suppose $f(X) = G(X)H(X)$ where $G(X), H(X) \in K[X]$. Set $G(X) = g(X)/d$ and $H(X) = h(X)/e$ where d and e are the least common denominators of the coefficients of G and H, respectively, and $g(X), h(X) \in R[X]$. Set $p(X) = g(X)/c(g)$ so that $p(X)$ is a primitive polynomial in $R[X]$. Then $de\, f(X) = c(g)h(X)p(X)$. But $R[X]$ is a unique factorization domain, primes in R are primes in $R[X]$, and $p(X)$ is primitive. Therefore de divides $c(g)h(X)$ so $f(X)$ factors over $R[X]$.

EXERCISES

4-1. Let R be a Euclidean domain, $a \in R$. Prove that a

is a unit in R if and only if d(a) = d(1).

4-2. Prove that every prime ideal in a Euclidean domain is maximal. Show by example, that this is false for unique factorization domains.

4-3. Define $d : \mathbb{Z}[i] \to \mathbb{Z}$ by $d(a+bi) = a^2 + b^2$. Prove that this function gives $\mathbb{Z}[i]$ the structure of a Euclidean domain.

4-4. (Factor Theorem) Let k be a field, $a \in k$ a root of $f(X) = 0$ where $f(X) \in k[X]$. Prove that X - a divides f(X).

4-5. Prove that $\mathbb{Z}[X]$ is not a principal ideal domain.

4-6. Prove that in a principal ideal domain, every ideal is a unique product of prime ideals.

4-7. Prove the remark following Theorem 4-7, that is, every unique factorization domain satisfies the ascending chain condition on principal ideals.

4-8. (Eisenstein's Criterion) Let $f(X) = a_0 + a_1 X + \cdots + a_n X^n \in \mathbb{Z}[X]$ and suppose p is a prime number such that p divides

a_i for $i = 0, 1, \ldots, n-1$, p does not divide a_n, and p^2 does not divide a_0. Prove that f(X) is irreducible in $\mathbb{Q}[X]$.

4-9. Let p be a prime number. Prove that $f(X) = 1 + X + X^2 + \cdots + X^{p-1} = X^p - 1/X - 1$ is irreducible in $\mathbb{Q}[X]$. (Hint: If f(X) factors, so does f(X+1). Substitute X+1 for X and apply Eisenstein's Criterion.)

4-10. A ring Λ is called <u>regular</u> if for any $a \in \Lambda$, there exists $b \in \Lambda$ such that $aba = a$. Suppose that Λ is regular. Prove each of the following:

(a) Every non-zero divisor of Λ is a unit.

(b) Every prime ideal of Λ is maximal.

(c) Every principal left ideal of Λ is generated by an element e satisfying $e^2 = e$.

CHAPTER 5

MODULES AND EXACT SEQUENCES

DEFINITION 5-1. Let Λ be a ring (not necessarily commutative). An Abelian group $(M, +)$ is called a <u>left Λ-module</u> if there is a map $\Lambda \times M \to M$ given by $(\lambda, m) \to \lambda m$ satisfying

(1) $\lambda(x+y) = \lambda x + \lambda y$ for any $\lambda \in \Lambda$, $x, y \in M$;

(2) $(\lambda+\mu)x = \lambda x + \mu x$ for any $\lambda, \mu \in \Lambda$, $x \in M$;

(3) $\lambda(\mu x) = (\lambda\mu)x$ for any $\lambda, \mu \in \Lambda$, $x \in M$;

(4) $1x = x$ for all $x \in M$.

A <u>right Λ-module</u> is similarly defined.

<u>Note</u>. If Λ is commutative, then every left Λ-module is a right Λ-module and vice versa. For if M is a left Λ-module, we make it a right Λ-module by defining $x\lambda = \lambda x$ for all $\lambda \in \Lambda$, $x \in M$. Then $(x\lambda)\mu = \mu(x\lambda) = \mu(\lambda x) = (\mu\lambda)x = x(\mu\lambda)$ $= x(\lambda\mu)$ where in the last step we use the commutativity of Λ.

In this situation we will just refer to M as a Λ-module.

DEFINITION 5-2. If M is a left Λ-module, then a nonempty subset N of M is called a *left Λ-submodule* if

(1) $x, y \in N$ implies $x - y \in N$;

(2) $x \in N$, $\lambda \in \lambda$ implies $\lambda x \in N$.

If N is a left Λ-submodule of M, we define an equivalence relation on M by $x \sim y$ if and only if $x - y \in N$. Denote by M/N the set of distinct equivalence classes. Then M/N is a left Λ-module under the operations $\overline{x} + \overline{y} = \overline{x + y}$ and $\lambda\overline{x} = \overline{\lambda x}$. It is called the *factor module of M by N* and is read "M mod N".

DEFINITION 5-3. Let M_i, $i \in I$ (finite or infinite), be left Λ-modules. Then the *direct sum*, denoted $\bigoplus_{i \in I} M_i$, is equal to

$$\{(\ldots, m_i, \ldots) \mid m_i \in M_i \text{ provided } m_i = 0_i \text{ for all but a finite number of } i \in I\}.$$

Elements of $\bigoplus_{i \in I} M_i$ are added coordinatewise and $\lambda(\ldots, m_i, \ldots) = (\ldots, \lambda m_i, \ldots)$. This makes $\bigoplus_{i \in I} M_i$ into a left Λ-module. Each M_i is called a *direct summand*.

DEFINITION 5-4. If $(M, +)$ and $M', +')$ are left Λ-modules and $f : M \to M'$, f is called a Λ-homomorphism if for any $x, y \in M$, $\lambda \in \Lambda$, $f(x+y) = f(x) +' f(y)$ and $f(\lambda x) = \lambda f(x)$.

The terms Λ-monomorphism, Λ-epimorphism, Λ-isomorphism have the obvious meaning. Also $\text{im } f = \{y \in M' \mid y = f(x) \text{ for some } x \in M\}$ and $\ker f = \{x \in M \mid f(x) = 0\}$. Observe that ker f is a left Λ-submodule of M and im f is a left Λ-submodule of M'.

THEOREM 5-1. Let $f : M \to M'$ be a Λ-homomorphism with im f = N and ker f = K. Then $N \approx M/K$. Furthermore there is a bijection between left Λ-submodules of M which contain K and left Λ-submodules of N.

Proof. The proof is virtually identical with those of the First and Second Isomorphism Theorems for rings and is left as an exercise.

DEFINITION 5-5. A set of elements X in M is said to generate M if for any element $m \in M$,

$$m = \lambda_1 x_1 + \lambda_2 x_2 + \cdots + \lambda_t x_t$$

for some $\lambda_1, \ldots \lambda_t \in \Lambda$, $x_1, \ldots, x_t \in X$. M is said to be finitely generated if there exists a finite subset X of M which generates M. The elements of such a set are called generators of M.

DEFINITION 5-6. A set of elements X in M is called a basis for M if

(1) X generates M;

(2) If $\Sigma_{i=1}^{t} \lambda_i x_i = 0$, then $\lambda_1 = \lambda_2 = \cdots = \lambda_t = 0$ for $\lambda_i \in \Lambda$, $x_i \in X$. In other words, X is a linearly independent set of elements.

THEOREM 5-2. Let M be a left Λ-module. The following statements are equivalent:

(1) $M \approx \bigoplus_{i \in I} \Lambda_i$, where $\Lambda_i = \Lambda$, I, any index set;

(2) M admits a basis.

Proof. (1) implies (2). We have that $(1, 0, \ldots, 0, \ldots)$, $(0, 1, 0, \ldots, 0, \ldots)$, and so forth form a basis for $\bigoplus_{i \in I} \Lambda_i$. Since isomorphisms carry bases to bases M has a basis.

(2) implies (1). Let $\{m_i\}_{i \in I}$ be a basis for M. Then $M = \bigoplus_{i \in I} \Lambda m_i$. The sum is direct by condition (2) on a basis. It suffices to show that $\Lambda m_i \approx \Lambda$ for each i. Consider the map $\Lambda \to \Lambda m_i$ given by $\lambda \to \lambda m_i$. This is a Λ-homomorphism and is clearly surjective. If $\lambda m_i = 0$, $\lambda = 0$ by condition (2) of a basis. Hence the map is injective,

hence a Λ-isomorphism. Therefore $M \approx \bigoplus_{i \in I} \Lambda_i$ where $\Lambda_i = \Lambda$.

DEFINITION 5-7. A left Λ-module M satisfying one (hence both) of the conditions of the above theorem is called a free left Λ-module.

EXAMPLES. 1. Any vector space V over a field k, is a free k-module.

2. Any Abelian group is a $\mathbb{Z}$-module. $\mathbb{Z}_2$ is not free over $\mathbb{Z}$.

3. Any ring Λ is a free Λ-module.

4. Any left ideal A of a ring Λ is a left Λ-module.

DEFINITION 5-8. A sequence of left Λ-modules and Λ-homomorphisms $M_1 \xrightarrow{f} M_2 \xrightarrow{g} M_3$ is called exact if ker g = im f. A sequence $0 \to M_1 \xrightarrow{f} M_2 \xrightarrow{g} M_3 \to 0$ is exact if

(1) ker f = 0, that is, f is injective;

(2) ker g = im f;

(3) im g = M_3, that is, g is surjective.

Such a configuration is called a short exact sequence of left Λ-modules and Λ-homomorphisms.

EXAMPLE. Suppose N is a left Λ-submodule of M. Then $0 \to N \overset{i}{\to} M \overset{j}{\to} M/N \to 0$ is a short exact sequence where i is the inclusion map and j is the canonical map $x \to \bar{x}$.

THEOREM 5-3. Let $0 \to M' \overset{f_1}{\to} M \overset{f_2}{\to} M'' \to 0$ be exact. The following statements are equivalent:

(1) $M \approx M' \oplus M''$.

(2) There exists a Λ-homomorphism $g_1 : M \to M'$ such that $g_1 f_1 = I_{M'}$, the identity map on M'.

(3) There exists a Λ-homomorphism $g_2 : M'' \to M$ such that $f_2 g_2 = I_{M''}$, the identity map on M''.

Proof. (1) implies (2). Since $M \approx M' \oplus M'' \approx f_1(M') \oplus M''$ there is a projection map $\varphi_1 : M \to f_1(M')$. Define $g_1 : M \to M'$ by $g_1 = f_1^{-1} \circ \varphi_1$. Then $g_1 f_1 = I_{M'}$.

(2) implies (3). Given $x'' \in M''$ choose $x \in M$ such that $f_2(x) = x''$. Define $g_2(x'') = x - f_1 g_1(x) \in M$. Thus $g_2 : M'' \to M$. To see that g_2 is well-defined, suppose also $y \in M$ such that $f_2(y) = x''$. Then $f_2(x - y) = x'' - x'' = 0$ so $x - y \in \ker f_2 = \operatorname{im} f_1$, say $x - y = f_1(x')$ for some $x' \in M'$. Then $x - f_1 g_1(x) = y + f_1(x') - f_1 g_1(y + f_1(x')) = y + f_1(x') - f_1 g_1(y) - f_1(x') = y - f_1 g_1(y)$. Therefore g_2 is well-defined. Finall $f_2 g_2(x'') = f_2(x) - f_2 f_1 g_1(x) = f_2(x) = x''$ since $f_2 f_1 = 0$.

(3) implies (1). If $x \in M$, then we can write

$x = g_2f_2(x) + (x - g_2f_2(x))$. Since $f_2(x - g_2f_2(x)) = f_2(x) - f_2(x) = 0$, $x - g_2f_2(x) \in \ker f_2 = \operatorname{im} f_1$. Since $f_2(x) \in M''$, we can say that $M = g_2(M'') + f_1(M')$. We must show this sum is direct, that is, $g_2(M'') \cap f_1(M') = 0$. Suppose $z \in g_2(M'') \cap f_1(M')$. Then $z = f_1(z') = g_2(z'')$ for some $z' \in M'$, $z'' \in M''$. Now $0 = f_2f_1(z') = f_2(z) = f_2g_2(z'') = z''$ so $z'' = 0$. Therefore $z = g_2(z'') = g_2(0) = 0$. So $M = g_2(M'') \oplus f_1(M')$. But $M' \approx f_1(M')$ and $M'' \approx g_2(M'')$. Thus $M \approx M' \oplus M''$.

DEFINITION 5-9. An exact sequence satisfying one (hence all) of the above conditions is called split (or is said to split). The maps g_1 and g_2 are called a retraction and a cross-section (section), respectively.

THEOREM 5-4. Let P be a left Λ-module. The following statements are equivalent:

(1) P is a direct summand of a free left Λ-module.

(2) Given any diagram

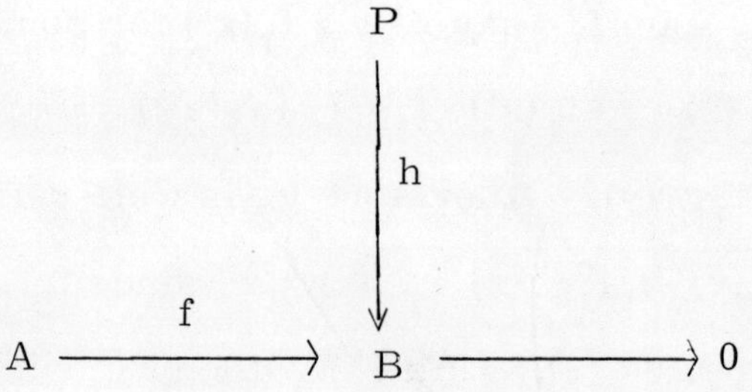

there exists a Λ-homomorphism $g : P \to A$ such that $fg = h$.

(3) Every exact sequence $0 \to M' \to M \to P \to 0$ splits.

Proof. (1) implies (2). First consider the diagram

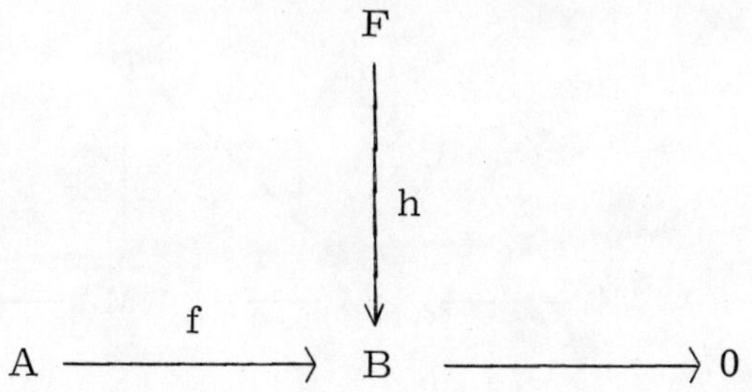

where F is a free Λ-module. If x is a basis element of F, define $g(x) = a$ for some $a \in A$ satisfying $f(a) = h(x)$. Extend by linearity to all of F giving $g : F \to A$ such that $fg = h$. Thus we can always "fill in" diagrams when F is free.

Now consider the diagram

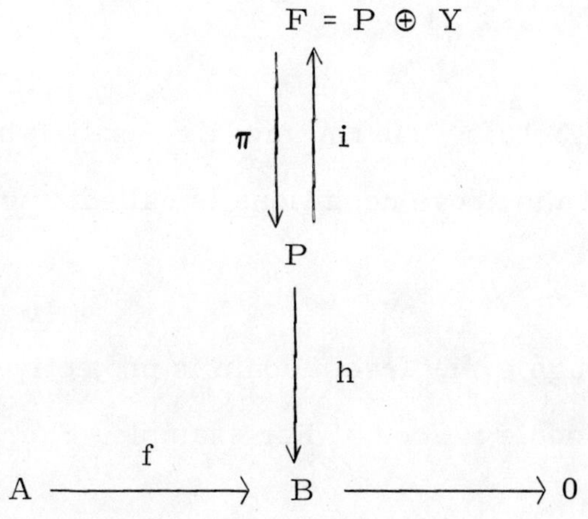

where F is free, P a direct summand of F, $i : P \to F$ is given by $x \to (x, 0)$ and $\pi : F \to P$ is given by $(x, y) \to x$. By the above there exists $\nu : F \to A$ such that $f\nu = h\pi$. Define $g : P \to A$ by $g = \nu i$. Then $fg = f\nu i = h\pi i = h$ since $\pi i = I_P$.

(2) implies (3). Consider the diagram

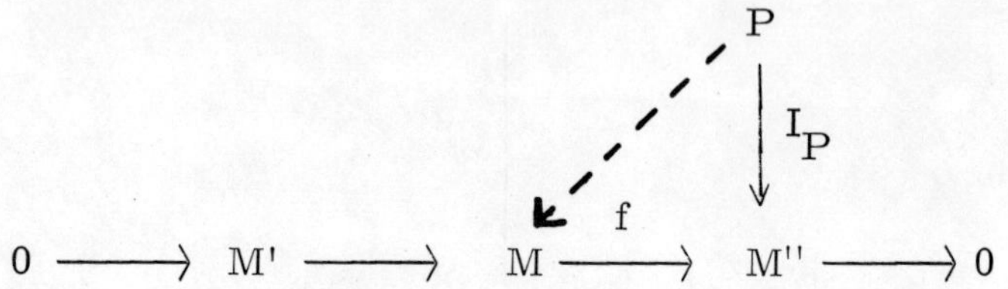

By (2), there exists $g : P \to M$ such that $fg = I_P$. By Theorem 5-3, the sequence splits.

(3) implies (1). Consider the exact sequence $0 \to \ker \nu \to F_\Lambda(P) \overset{\nu}{\to} P \to 0$ where $F_\Lambda(P) = \oplus_{\alpha \in P} \Lambda_\alpha$, $\Lambda_\alpha = \Lambda$ for all α and $\nu(\ldots, \lambda_\alpha, \ldots) = \Sigma_{\alpha \in P} \lambda_\alpha \alpha$ in P. Since this sequence splits, $F_\Lambda(P) \approx P \oplus \ker \nu$ so P is a direct summand of a free module.

DEFINITION 5-10. A left Λ-module P satisfying one (hence all) of the above conditions is called a projective left Λ-module.

Note. Although every free module is projective, not every projective module is free. For example, Exercise 2-10 states that $\mathbb{Z}_6$ and $\mathbb{Z}_2 \oplus \mathbb{Z}_3$ are ring isomorphic, hence isomorphic as $\mathbb{Z}_6$-modules. Thus $\mathbb{Z}_2$ is a projective but not a free $\mathbb{Z}_6$-module.

Suppose now R is a commutative ring with unit, S is a multiplicative set in R, and M is an R-module. Then we can define a <u>module of quotients</u> M_S which is a module over the ring of quotients R_S as follows:

In the set $\{m/s \mid m \in M, s \in S\}$, we define an equivalence relation by $m_1/s_1 \sim m_2/s_2$ if there exists $s \in S$ such that $s(s_1 m_2 - s_2 m_1) = 0$. M_S is the set of equivalence classes. We define + in M_S by

$$[m_1/s_1] + [m_2/s_2] = [s_2 m_1 + s_1 m_2/s_1 s_2]$$

and if $[r/s] \in R_S$,

$$[r/s] \cdot [m_1/s_1] = [rm_1/ss_1].$$

In this way M_S is an R_S-module. If $S = R - P$ for some prime ideal P of R, we denote M_S by M_P. Finally there is an R-homomorphism $M \to M_S$ given by $m \to [m/1]$.

We conclude this chapter with a pair of isomorphism theorems for modules which will be particularly useful in Chapter 8.

THEOREM 5-5. Let M be a left Λ-module with M_1 and M_2 left Λ-submodules of M. Then $M_1/M_1 \cap M_2$ is isomorphic to $M_1 + M_2/M_2$.

<u>Proof</u>. Consider the Λ-homomorphism $\sigma : M_1 \to M_1 + M_2/M_2$ given by composing the inclusion map of M_1 into $M_1 + M_2$ with the canonical surjection from $M_1 + M_2$ onto $M_1 + M_2/M_2$. Then σ is surjective and $\ker \sigma = M_1 \cap M_2$ so by Theorem 5-1, the result follows.

LEMMA. Let M_1, M_2, M_3 be left Λ-submodules of a left Λ-module M. Suppose $M_3 \subseteq M_1$. Then $M_1 \cap (M_2 + M_3) = (M_1 \cap M_2) + M_3$.

<u>Proof</u>. If $x \in M_1 \cap M_2$, then $x \in M_1$ and $x \in M_2 + M_3$ so $x \in M_1 \cap (M_2 + M_3)$. Also if $x \in M_3 \subseteq M_1$, then $x \in M_1 \cap (M_2 + M_3)$. Thus $M_1 \cap (M_2 + M_3) \supseteq (M_1 \cap M_2) + M_3$.

Suppose then that $x \in M_1 \cap (M_2 + M_3)$, that is, $x = y + z$ where $y \in M_2$, $z \in M_3 \subseteq M_1$. Then $y = x - z \in M_1$ so $y \in M_1 \cap M_2$. Thus $x \in (M_1 \cap M_2) + M_3$. Therefore $M_1 \cap (M_2 + M_3) \subseteq (M_1 \cap M_2) + M_3$, completing the proof.

THEOREM 5-6. Let $M_1' \subseteq M_1$, $M_2' \subseteq M_2$, be left Λ-submodules of a left Λ-module M. Then $M_1' + (M_1 \cap M_2)/M_1' + (M_1 \cap M_2')$ is isomorphic to

$M_2' + (M_1 \cap M_2)/M_2' + (M_1' \cap M_2)$.

Proof. Let $N = M_1 \cap M_2$, $N_1 = M_1' + (M_1 \cap M_2')$, and $N_2 = M_2' + (M_1' \cap M_2)$. Thus $N + N_1 = M_1' + (M_1 \cap M_2)$ and $N + N_2 = M_2' + (M_1 \cap M_2)$. We wish to prove $N + N_1/N_1$ is isomorphic to $N + N_2/N_2$. But according to Theorem 5-5, this is equivalent to showing $N/N \cap N_1$ is isomorphic to $N/N \cap N_2$. Finally by the previous lemma, we have $N \cap N_1 = (M_1' \cap M_2) + (M_1 \cap M_2') = N \cap N_2$, completing the proof.

EXERCISES

5-1. Let A and B be left Λ-modules and denote by $\mathrm{Hom}_\Lambda(A, B)$ the set of all Λ-homomorphisms from A to B. If f, g $\in \mathrm{Hom}_\Lambda(A, B)$, $x \in A$, $\lambda \in \Lambda$, define f + g and λf by $(f+g)(x) = f(x) + g(x)$ and $(\lambda f)(x) = \lambda f(x)$. Show that these operations make $\mathrm{Hom}_\Lambda(A, B)$ a left Λ-module.

5-2. Prove Theorem 5-1.

5-3. Suppose that a left Λ-module M has no left Λ-submodule other than 0 and M. Prove that $M \approx \Lambda/\mathfrak{m}$ for some maximal left ideal $\mathfrak{m}$ of Λ.

5-4. Let R be a commutative ring and M an R-module. Suppose that $M_P = 0$ for every maximal ideal P of R. Prove that $M = 0$.

5-5. Let R be a commutative ring, S a multiplicative set in R, and $0 \to A \overset{f}{\to} B \overset{g}{\to} C \to 0$ an exact sequence of R-modules. Define $\tilde{f} : A_S \to B_S$ by $\tilde{f}([a/s]) = [f(a)/s]$ and $\tilde{g} : B_S \to C_S$ by $\tilde{g}([b/s]) = [g(b)/s]$. Show that $\tilde{f}$ and $\tilde{g}$ are well-defined R_S-homomorphisms and that $0 \to A_S \overset{\tilde{f}}{\to} B_S \overset{\tilde{g}}{\to} C_S \to 0$ is an exact sequence of R_S-modules.

5-6. (Five lemma) Consider the following diagram of left Λ-modules and Λ-homomorphisms:

$$\begin{array}{ccccccccc} 0 & \longrightarrow & A & \xrightarrow{f} & B & \xrightarrow{g} & C & \longrightarrow & 0 \\ & & \downarrow{\scriptstyle\alpha} & & \downarrow{\scriptstyle\beta} & & \downarrow{\scriptstyle\gamma} & & \\ 0 & \longrightarrow & A' & \xrightarrow{f'} & B' & \xrightarrow{g'} & C' & \longrightarrow & 0 \end{array}$$

where each row is an exact sequence, $f'\alpha = \beta f$ and $g'\beta = \gamma g$. Prove each of the following:

(1) If β is a monomorphism, then α is a monomorphism.

(2) If β is an epimorphism, then γ is an epimorphism.

(3) If α and γ are monomorphisms (epimorphisms), then β is a monomorphism (epimorphism).

(4) If β is an isomorphism, then $\ker \gamma \approx A'/\mathrm{im}\, \alpha$.

5-7. (Snake diagram) If $f : A \to B$ is a homomorphism of left Λ-modules, then the cokernel of f, denoted coker f, is equal to the factor module $B/\mathrm{im}\, f$. Given the diagram in Exercise 4-7, prove that there exists an exact sequence
$\ker \alpha \to \ker \beta \to \ker \gamma \to \mathrm{coker}\, \alpha \to \mathrm{coker}\, \beta \to \mathrm{coker}\, \gamma$.

5-8. Let C be the set of all real-valued functions of two real variables which are infinitely differentiable. Then C can be viewed as an $\mathbb{R}$-module under the operations $(f+g)(x, y) = f(x, y) + g(x, y)$ and $(\alpha f)(x, y) = \alpha f(x, y)$ for any $\alpha \in \mathbb{R}$. Let $C_0 = \{f \in C \mid f(0, 0) = 0\}$. Define maps $\varphi : C_0 \to C \oplus C$ and $\psi : C \oplus C \to C$ by $\varphi(f) = (f_x, f_y)$ and $\psi(g, h) = g_y - h_x$. Prove:

(a) φ and ψ are $\mathbb{R}$-homomorphisms.

(b) The sequence $0 \to C_0 \xrightarrow{\varphi} C \oplus C \xrightarrow{\psi} C \to 0$ is exact.

(Hint: Recall the notion of an exact differential from calculus.)

5-9. Let R be a principal ideal domain. Prove that every ideal of R is a free R-module.

5-10. Prove that a ring Λ is regular if and only if every principal left ideal of Λ is a projective Λ-module. (See Exercise 4-10.)

CHAPTER 6

NOETHERIAN RINGS AND MODULES

In a unique factorization domain, one can write any nonunit $a = p_1^{\alpha_1} p_2^{\alpha_2} \cdots p_t^{\alpha_t}$ where $p_1, p_2, \ldots, p_t$ are distinct primes and $\alpha_i \geq 1$ for $i = 1, \ldots, t$. Furthermore this factorization is unique up to order and units. Viewing this result in terms of ideals we see that

$$(a) = (p_1^{\alpha_1})(p_2^{\alpha_2}) \cdots (p_t^{\alpha_t}) = (p_1^{\alpha_1}) \cap (p_2^{\alpha_2}) \cap \cdots \cap (p_t^{\alpha_t}).$$

In this chapter we study a class of commutative rings for which a similar factorization exists for all ideals. In the course of the discussion, we will see that the factorization will be as an intersection rather than a product of ideals. Also the factors will not be simply powers of prime ideals but things called primary ideals. Finally there will be the question of the uniqueness of the factorization. We will find that the uniqueness is only partial, in a sense to be explained later.

THEOREM 6-1. Let Λ be a ring (not necessarily commutative) and M a left Λ-module. The following conditions are equivalent:

(1) Every strictly ascending chain of left Λ-submodules of M, $M_1 \subset M_2 \subset M_3 \subset \cdots$ is finite.

(2) Every nonempty collection of left Λ-submodules of M has a maximal element.

(3) Every left Λ-submodule of M is finitely generated.

Note. Since M is a submodule of itself, (3) implies M is finitely generated.

Proof. The equivalence of (1) and (2) is proven the same way it was done earlier for rings and ideals. Let us introduce the notation $\langle x_1, \ldots, x_n \rangle = \{\lambda_1 x_1 + \cdots + \lambda_n x_n \mid \lambda_i \in \Lambda\}$ to denote the left Λ-module generated by $x_1, \ldots, x_n \in M$. We will show that (1) and (3) are equivalent.

(3) implies (1). Suppose $M_1 \subseteq M_2 \subseteq M_3 \subseteq \cdots$ is a chain of left Λ-submodules of M. Let $N = \cup_i M_i$. Then N is a left Λ-submodule of M and so is finitely generated, say, $N = \langle x_1, \ldots, x_n \rangle$. Then each $x_i \in M_j$ for some $j = j(i)$. Let $k = \max_{i=1,\ldots,n} j(i)$. Then $x_i \in M_k$ for all i.

Thus $\langle x_1, \ldots x_n \rangle \subseteq M_k \subseteq N = \langle x_1, \ldots, x_n \rangle$ so $N = M_k$. If $r \geq k$, $M_k \subseteq M_r \subseteq N$ implies $M_r = M_k$. Thus the chain is stable, that is, all the inclusions are equalities beyond some point.

(1) implies (3). Let N be a left Λ-submodule of M and $x_1 \in N$. If $N = \langle x_1 \rangle$, we are done. Otherwise there exists $x_2 \in N$, $x_2 \notin \langle x_1 \rangle$, whence $\langle x_1 \rangle \subset \langle x_1, x_2 \rangle$. If $N = \langle x_1, x_2 \rangle$, we are done. Otherwise choose x_3, and so forth. Since $\langle x_1 \rangle \subset \langle x_1, x_2 \rangle \subset \langle x_1, x_2, x_3 \rangle \subset \cdots$ is finite, we get for some integer n, $N = \langle x_1, x_2, \ldots, x_n \rangle$. Thus N is finitely generated.

DEFINITION 6-1. A left Λ-module M satisfying any (hence all) of the above conditions is called <u>left Noetherian</u>. A ring Λ is called left Noetherian if it is a left Noetherian Λ-module. A commutative ring R is called Noetherian if it is Noetherian as an R-module.

THEOREM 6-2. Let $0 \to M' \to M \xrightarrow{f} M'' \to 0$ be an exact sequence of left Λ-modules. Then M is left Noetherian if and only if M' and M'' are left Noetherian.

<u>Proof</u>. (a) Suppose M is left Noetherian. Since every submodule of M' is a submodule of M, M' is left Noetherian.

Let $M_1'' \subseteq M_2'' \subseteq \cdots$ be a chain of submodules of M''. Then setting $M_i = f^{-1}(M_i'')$, we have a chain $M_1 \subseteq M_2 \subseteq \cdots$ of submodules of M. Thus there exists an integer k such that $M_i = M_k$ for $i \geq k$. Therefore $M_i'' = M_k''$ for all $i \geq k$ so M'' is Noetherian.

(b) Suppose $M_1 \subseteq M_2 \subseteq \cdots$ is a chain of submodules of M. Set $M_i' = M_i \cap M'$ and $M_i'' = M_i + M'/M'$. Then $M_1' \subseteq M_2' \subseteq \cdots$ is a chain of submodules of M' and $M_1'' \subseteq M_2'' \subseteq \cdots$ is a chain of submodules of M''. Since M' and M'' are Noetherian, there exist integers r and s such that $M_i' = M_r'$ for all $i \geq r$ and $M_j'' = M_s''$ for all $j \geq s$. Let $t = \max(r, s)$. The claim is that $M_k = M_t$ for all $k \geq t$. Suppose $x \in M_k$. Since $M_k + M'/M' = M_t + M'/M'$, there exist $y \in M_t$ and $a \in M'$ such that $x - y = a$. Therefore $x - y \in M_k \cap M' = M_k' = M_t' \subseteq M_t$. Since $y \in M_t$, we conclude $x \in M_t$. Therefore $M_k \subseteq M_t$ whence $M_k = M_t$.

COROLLARY 1. If M is a left Noetherian module, every submodule and factor module of M is left Noetherian.

COROLLARY 2. A finite direct sum of left Noetherian modules is left Noetherian.

THEOREM 6-3. Suppose Λ is a left Noetherian ring.

Then a left Λ-module M is finitely generated if and only if M is left Noetherian.

Proof. Suppose $M = \langle x_1, \ldots, x_n \rangle$. Set $F = \bigoplus_{i=1}^{n} \Lambda_i$, $\Lambda_i = \Lambda$ for all i, the free left Λ-module of rank n. By the previous Corollary 2, F is left Noetherian. Also $f : F \to M$ given by $f(\lambda_1, \ldots, \lambda_n) = \Sigma_{i=1}^{n} \lambda_i x_i$ is an epimorphism. Thus M is a factor module of F and so is left Noetherian by Corollary 1. The other implication is true by definition.

THEOREM 6-4. If R is a commutative Noetherian ring and S a multiplicative subset of R, then R_S is Noetherian.

Proof. Let A be an ideal of R_S. Then $A \cap R$ is an ideal of R which is finitely generated. A basis for $A \cap R$ over R is in turn a basis for A over R_S, whence A is finitely generated.

THEOREM 6-5. (Hilbert Basis Theorem) If R is a commutative Noetherian ring, then R[X] is Noetherian.

Before proving this important theorem, we will need some preliminary lemmas.

DEFINITION 6-2. If A is any ideal in R[X], set

$$A_n = \{a \in R \mid a \text{ is the coefficient of } X^n \text{ in some}$$
$$f(X) \in A \text{ with } \deg f(X) \le n\}.$$

A_n is called the nth associated ideal of A.

LEMMA 6-1. If A is an ideal of R[X], then A_n is an ideal of R. Furthermore $A_n \subseteq A_{n+1}$ for all n.

Proof. If a, b $\in A_n$, say $f(X) = a_0 + a_1X + \cdots + aX^n$, $g(X) = b_0 + b_1X + \cdots + bX^n \in A$, then $f(X) - g(X) = (a_0 - b_0) + \cdots + (a - b)X^n \in A$ so $a - b \in A_n$. Also if $r \in R$, $rf(X) = ra_0 + ra_1X + \cdots + raX^n \in A$ so $ra \in A_n$. Thus A_n is an ideal. Finally $Xf(X) = a_0X + a_1X^2 + \cdots + aX^{n+1} \in A$ so $a \in A_{n+1}$. Therefore $A_n \subseteq A_{n+1}$.

LEMMA 6-2. Let $A \subseteq B$ be ideals of R[X]. Then $A_n \subseteq B_n$ for all n. Furthermore if $A_n = B_n$ for all n, then A = B.

Proof. The first statement is immediate from Definition 6-2. Suppose now that $A_n = B_n$ for all n, and that $f(X) = b_0 + b_1X + \cdots + b_nX^n \in B$. We wish to show $f(X) \in A$. The proof will be by induction on n.

If n = 0, then $f(X) = b_0 \in B_0 = A_0 \subseteq A$. Thus assume that the statement is true for polynomials of degree $\leq n-1$. Since $f(X) \in B$, we have $b_n \in B_n = A_n$. Thus there exists a polynomial $g(X) = a_0 + a_1 X + \cdots + a_{n-1} X^{n-1} + b_n X^n \in A \subseteq B$. Thus $f(X) - g(X) \in B$. But $\deg(f(X) - g(X)) \leq n - 1$ so by the induction hypothesis $f(X) - g(X) \in A$. Since $g(X) \in A$, we conclude $f(X) \in A$, completing the induction.

Proof (Theorem 6-5). Suppose $A_0 \subseteq A_1 \subseteq A_2 \subseteq \cdots$ is a chain of ideals of $R[X]$. Let $A_{i,j}$ denote the jth associated ideal of A_i. We then have the following pattern of inclusions:

$$\begin{array}{ccccccc}
\vdots & & \vdots & & \vdots & & \\
A_{2,0} & \subseteq & A_{2,1} & \subseteq & A_{2,2} & \subseteq & \cdots \\
\cup\!| & & \cup\!| & & \cup\!| & & \\
A_{1,0} & \subseteq & A_{1,1} & \subseteq & A_{1,2} & \subseteq & \cdots \\
\cup\!| & & \cup\!| & & \cup\!| & & \\
A_{0,0} & \subseteq & A_{0,1} & \subseteq & A_{0,2} & \subseteq & \cdots
\end{array}$$

Since $A_{0,0} \subseteq A_{1,1} \subseteq A_{2,2} \subseteq \cdots$ is an ascending chain of ideals of R, there exists an integer k such that $A_{i,i} = A_{k,k}$ for $i \geq k$. Now consider the k vertical chains $A_{0,j} \subseteq A_{1,j} \subseteq A_{2,j} \subseteq \cdots$, $j = 0, 1, \ldots, k-1$. There exists an integer n_j such that for all $t \geq n_j$, $A_{t,j} = A_{n_j,j}$. Let $n = \max\{n_0, n_j, \ldots, n_{k-1}, k\}$. The claim is that if $i \geq n$, $A_i = A_n$. By the lemma, it suffices to show

$A_{i,j} = A_{n,j}$ for all j. If $0 \leq j < k$, then $i \geq n \geq n_j$ implies $A_{i,j} = A_{n_j,j} = A_{n,j}$. If $j \geq k$ and $i \geq n \geq k$ then $A_{i,j} = A_{k,k} = A_{n,j}$ since $A_{r,s} = A_{k,k}$ whenever $r, s \geq k$. This completes the proof.

For the remainder of this chapter, R is a commutative Noetherian ring with unit element.

DEFINITION 6-3. Let M be an R-module. Then the annihilator of M, denoted (0:M), is equal to $\{r \in R \mid rx = 0 \text{ for all } x \in M\}$.

Certain facts are immediate from this definition. In particular (0:M) is an ideal in R and if $N \subseteq M$, then $(0:N) \supseteq (0:M)$.

THEOREM 6-6. Let M be an R-module. The following statements are equivalent:

(1) $(0:M') = (0:M)$ for all $0 \neq M' \subseteq M$.

(2) $AM' = 0$ implies $AM = 0$ for all $0 \neq M' \subseteq M$ and all ideals A of R.

Proof. (1) implies (2). $AM' = 0$ implies $A \subseteq (0:M') = (0:M)$ which implies $AM = 0$.

(2) implies (1). Take $A = (0:M')$. Then $(0:M')M' = 0$ so $(0:M')M = 0$ so $(0:M') \subseteq (0:M) \subseteq (0:M')$, the latter inclusion following from the above remark. Thus $(0:M') = (0:M)$.

DEFINITION 6-4. An R-module $M \neq 0$ satisfying one (hence both) of the above conditions is called a <u>prime module</u>.

THEOREM 6-7. (1) R/P is a prime module if and only if P is a prime ideal.

(2) If M is a prime module, then $(0:M)$ is a prime ideal.

<u>Proof</u>. (1) Suppose R/P is a prime module, $AB \subseteq P$ and $A \not\subseteq P$. Then $0 \neq A/P \subseteq R/P$. Then $B \subseteq (0:A/P) = (0:R/P) = P$ so P is a prime ideal. Conversely, suppose P is a prime ideal, A/P a nonzero submodule of R/P, and B any ideal of R. Then $B(A/P) = 0$ implies $BA \subseteq P$ which implies $B \subseteq P$ which implies $B(R/P) = 0$. Thus R/P is a prime module.

(2) Suppose M is a prime module and $AB \subseteq (0:M)$ and $B \not\subseteq (0:M)$. Then $0 \neq BM \subseteq M$. Also $A(BM) \subseteq (0:M)M = 0$ so $A \subseteq (0:BM) = (0:M)$. Thus $(0:M)$ is a prime ideal.

<u>Note</u>. The converse of (2) is false. For let $R = k[X, Y]$,

$M_1 = k[X, Y]/(X)$, $M_2 = k[X, Y]/(X, Y)$ and $M = M_1 \oplus M_2$. Then $(0:M) = (X)$ is a prime ideal but $(0:M_2) = (X, Y) \supset (X)$.

DEFINITION 6-5. Let M be an R-module. The associated primes of M, denoted A(M), is the set $\{P \mid P = (0:M') \text{ for some nonzero prime submodule } M' \text{ of } M\}$. This set is also called the assassinator of M but we will avoid this term.

DEFINITION 6-6. An R-module M is called a P-primary module if $A(M) = \{P\}$, that is, if P is its only associated prime. An ideal Q of R is called a P-primary ideal if $A(R/Q) = \{P\}$.

THEOREM 6-8. Let M be an R-module. Then

(1) $A(M) = \Phi$ if and only if $M = 0$.

(2) If $0 \to M' \to M \xrightarrow{f} M'' \to 0$ is exact, then
$A(M') \subseteq A(M) \subseteq A(M') \cup A(M'')$.

Proof. (1) Clearly $A(0) = \Phi$. On the other hand, let $M \neq 0$ and consider $\mathcal{S} = \{(0:M') \mid 0 \neq M' \subseteq M\}$. Since R is Noetherian $\mathcal{S}$ has a maximal element; call it $(0:M_1)$. Then if $0 \neq M' \subseteq M_1$, we have $(0:M') \supseteq (0:M_1)$ which implies $(0:M') = (0:M_1)$ by the maximality of $(0:M_1)$. Thus

M_1 is a prime submodule of M. Hence $A(M) \neq \Phi$.

(2) Clearly $A(M') \subseteq A(M)$. Let M_1 be a prime submodule of M and let $P = (0:M_1)$. If $M_1 \cap M' \neq 0$, then it is a prime submodule of M'. Thus $P \in A(M_1 \cap M') \subseteq A(M')$. Otherwise $M_1 \cap M' = 0$ which means f restricted to M_1 is injective. Thus $M_1 \approx f(M_1) \subseteq M''$. Thus $P \in A(f(M_1)) \subseteq A(M'')$. Hence $A(M) \subseteq A(M') \cup A(M'')$.

THEOREM 6-9. Let M be a Noetherian R-module and $A(M) = X \cup Y$ (union of disjoint sets). Then there exists an exact sequence $0 \to M' \to M \to M'' \to 0$ with $A(M') = X$ and $A(M'') = Y$.

Proof. Let $\mathcal{S} = \{M_1 \subseteq M \mid A(M_1) \subseteq X\}$. Then $\mathcal{S}$ is nonempty since 0 is in $\mathcal{S}$. By the maximal condition for Noetherian modules, there exists a maximal element M' in $\mathcal{S}$. Consider the exact sequence $0 \to M' \to M \xrightarrow{\nu} M'' \to 0$ where $M'' = M/M'$. Then $X \cup Y = A(M) \subseteq A(M') \cup A(M'') \subseteq X \cup A(M'')$ which implies $A(M'') \supseteq Y$ since $X \cap Y = \Phi$.

We next claim that $A(M'') = Y$. Suppose $P \in A(M'')$ but $P \notin Y$. Let $0 \neq M''_2 \subseteq M''$ be a P-prime submodule of M''. Set $M_2 = \nu^{-1}(M''_2) \subseteq M$. We have an exact sequence $0 \to M' \to M_2 \to M''_2 \to 0$. Since $M''_2 \neq 0$, we know $M_2 \neq M'$. But $A(M_2) \subseteq A(M') \cup A(M''_2)$ so we then obtain

$A(M_2) \subseteq [A(M') \cup A(M_2'')] \cap [X \cup Y] \subseteq [X \cup \{P\}] \cap [X \cup Y] = X$ regardless of whether $P \in X$ or $P \notin X$. This contradicts the maximality of M', verifying the claim.

It now follows that $A(M') = X$. For $X \cup Y = A(M) \subseteq A(M') \cup A(M'') = A(M') \cup Y$ so $A(M') \supseteq X$. But by definition of $\mathcal{S}$, $A(M') \subseteq X$ so $A(M') = X$.

THEOREM 6-10. If M is a Noetherian R-module, A(M) is finite.

<u>Proof</u>. Suppose A(M) is infinite containing $P_1, P_2, P_3, \ldots$. Then we can find $M_1 \subset M_2 \subset M_3 \subset \cdots$ contradicting the ascending chain condition on M as follows: Clearly we can find $M_1 \subset M$ with $A(M_1) = P_1$. Then $0 \to M_1 \to M \to M/M_1 \to 0$ gives $P_2 \in A(M) \subseteq A(M_1) \cup A(M/M_1) = \{P_1\} \cup A(M/M_1)$ so $P_2 \in A(M/M_1)$. Thus there exists a submodule M_2 such that $0 \neq M_2/M_1 \subset M/M_1$ with $A(M_2/M_1) = \{P_2\}$. Now $0 \to M_1 \to M_2 \to M_2/M_1 \to 0$ gives $A(M_2) \subseteq A(M_1) \cup A(M_2/M_1) = \{P_1, P_2\}$. Now considering the exact sequence $0 \to M_2 \to M \to M/M_2 \to 0$, we get that $P_3 \in A(M) \subseteq A(M_2) \cup A(M/M_2) \subseteq \{P_1, P_2\} \cup A(M/M_2)$ so $P_3 \in A(M/M_2)$. Thus there exists a submodule M_3 such that $0 \neq M_3/M_2 \subset M/M_2$ such that $A(M_3/M_2) = \{P_3\}$. The construction and hence the contradiction should now be

evident.

DEFINITION 6-7. Let M be a Noetherian R-module. Then $0 = \cap_{i=1}^{t} M_i$, $M_i \subseteq M$, is called a <u>primary decomposition of 0 in M</u> if

(1) M/M_i is a P_i-primary module for $i = 1, \ldots, t$;

(2) $P_i \neq P_j$ for $i \neq j$;

(3) $M_1 \cap \cdots \cap \hat{M}_i \cap \cdots \cap M_t \neq 0$ for any $i = 1, \ldots, t$, where the symbol $\wedge$ means "omit the term below".

THEOREM 6-11. (1) Every Noetherian R-module M possesses a primary decomposition of 0 in M.

(2) If $0 = \cap_{i=1}^{t} M_i$ is a primary decomposition of 0 in M, then $A(M) = \cup_{i=1}^{t} A(M/M_i)$.

Proof. Let $A(M) = \{P_1, P_2, \ldots, P_t\}$. Choose $M_i \subset M$ such that $A(M_i) = A(M) - \{P_i\}$ and $A(M/M_i) = \{P_i\}$. Then $A(\cap_{i=1}^{t} M_i) \subseteq \cap_{i=1}^{t} A(M_i) = \Phi$ so $\cap_{i=1}^{t} M_i = 0$.

Suppose $M_1 \cap \cdots \cap \hat{M}_i \cap \cdots \cap M_t = 0$ for some i. Then $M \to M/M_1 \oplus \cdots \oplus \widehat{M/M_i} \oplus \cdots \oplus M/M_t$ is injective so $A(M) \subseteq A(M/M_1) \cup \cdots \cup A(\widehat{M/M_i}) \cup \cdots \cup A(M/M_t)$ $= \{P_1, \ldots, \hat{P}_i, \ldots, P_t\}$, contradiction. So we have exhibited a primary decomposition of 0 in M.

Now suppose $0 = \cap_{i=1}^{t} M_i$ is a primary decomposition of 0 in M. Then $M \to M/M_1 \oplus \cdots \oplus M/M_t$ is injective so $A(M) \subseteq \cup_{i=1}^{t} A(M/M_i)$. Furthermore for any i, $0 \neq M_1 \cap \cdots \cap \hat{M}_i \cap \cdots \cap M_t \to M/M_i$ is injective. Therefore $\Phi \neq A(M_1 \cap \cdots \cap \hat{M}_i \cap \cdots \cap M_t) \subseteq A(M/M_i) = \{P_i\}$ so $A(M_1 \cap \cdots \cap \hat{M}_i \cap \cdots \cap M_t) = \{P_i\}$. Therefore $\cup_{i=1}^{t} A(M/M_i) = \cup_{i=1}^{t} A(M_1 \cap \cdots \cap \hat{M}_i \cap \cdots \cap M_t) \subseteq A(M)$. Therefore $A(M) = \cup_{i=1}^{t} A(M/M_i)$.

THEOREM 6-12. Let A be an ideal in a Noetherian ring R. The following ideals are equal:

(1) $T_1 = \{x \in R \mid x^n \in A \text{ for some integer } n \geq 1\}$.

(2) T_2, the intersection of all prime ideals of R which contain A.

Proof. If $x \in T_1$, then $x^n \in A$ for some $n \geq 1$ so $x^n \in P$ for every prime ideal $P \supseteq A$ so $x \in P$ for every prime $P \supseteq A$ so $x \in T_2$.

If $x \notin T_1$, then $S = \{1, x, x^2, \ldots\}$ is a multiplicative set in R which does not meet A. We will show that there is a prime ideal of R which contains A and does not meet S. This will give $x \notin T_2$.

Set $\mathfrak{S} = \{B \mid B \supseteq A, \text{ B is an ideal of R}, B \cap S = \Phi\}$. Since A is in $\mathfrak{S}$, $\mathfrak{S}$ is nonempty and so has a maximal element P. Suppose $ab \in P$, $a \notin P$, $b \notin P$. Then by the maximality (a, P) and (b, P) meet S, that is, there exists $s_1, s_2, \in S$, $c_1, c_2 \in R$, $p_1, p_2 \in P$ such that $s_1 = c_1 a + p_1$ and $s_2 = c_2 b + p_2$. Then $s_1 s_2 = c_1 c_2 ab + c_1 ap_2 + c_2 bp_1 + p_1 p_2 \in P \cap S = \Phi$. Contradiction. So P is prime, $x \notin P$ and so $x \notin T_2$.

DEFINITION 6-8. T_1 (or T_2) in the above lemma is called the nil radical of A and is denoted $\sqrt{A}$.

THEOREM 6-13. For some integer N, $(\sqrt{A})^N \subseteq A$.

Proof. Since R is Neotherian, let $\sqrt{A} = (x_1, \ldots, x_t)$. Suppose $x_i^{n_i} \in A$. Set $N = \Sigma_{i=1}^t n_i$.

THEOREM 6-14. Let M be a Noetherian R-module. Then $\sqrt{(0:M)} = \cap_{P \in A(M)} P$.

Proof. $P \in A(M)$ implies $P = (0:M')$ for some prime submodule M' of M which implies $P \supseteq (0:M)$ which implies $P \supseteq \sqrt{(0:M)}$. Therefore $\cap_{P \in A(M)} P \supseteq \sqrt{(0:M)}$.

For any element $a \in R$, define an R-homomorphism $\tilde{a} : M \to M$ by $x \to ax$. Set $K_a = \ker \tilde{a} = \{x \in M \mid ax = 0\}$. We have $(\tilde{a})^k = \widetilde{a^k}$ and $K_a \subseteq K_{a^2} \subseteq \cdots$, an ascending chain of R-submodules of M.

Now suppose $r \in \bigcap_{P \in A(M)} P$. Then there exists an integer n such that $K_{r^n} = K_{r^{n+1}} = \cdots$. Set $N = r^n M$ and consider $\tilde{r} : N \to N$. If $x = r^n y \in N$ is in ker $\tilde{r}$, then $r^{n+1} y = 0$ so $y \in K_{r^{n+1}} = K_{r^n}$ so $x = r^n y = 0$. Thus $\tilde{r} : N \to N$ is injective. In fact, the module $N = 0$. For $N \neq 0$ if and only if $\Phi \neq A(N) \subseteq A(M)$ whence $r \in \bigcap_{P \in A(N)} P$. Thus $N \neq 0$ implies $r \in (0:N')$ for some $0 \neq N' \subseteq N$. Then $\tilde{r} : N \to N$ cannot be injective. Contradiction. Therefore $N = 0$ so $r^n \in (0:M)$ and $r \in \sqrt{(0:M)}$. Finally we have $\bigcap_{P \in A(M)} P = \sqrt{(0:M)}$.

THEOREM 6-15. The following properties of an ideal Q are equivalent:

(1) Q is a P-primary ideal, that is, $A(R/Q) = \{P\}$.

(2) If $ab \in Q$ and $b \notin Q$, then $a \in \sqrt{Q}$ for any $a, b \in R$.

Proof. (1) implies (2). Suppose $ab \in Q$, $b \notin Q$. Then $0 \neq bR/Q \subseteq R/Q$ and $a \in (0:bR/Q) \subseteq P$ since annihilators of all nonzero submodules of R/Q are contained in P. But $P = \bigcap_{\varphi \in A(R/Q)} \varphi = \sqrt{(0:R/Q)} = \sqrt{Q}$ by Theorem 6-14.

(2) implies (1). Set $P = \sqrt{Q}$. Then P is an ideal of R and $P \supseteq Q$. Furthermore P is a prime ideal of R. For if $ab \in P$, $a \notin P$, then $(ab)^n = a^n b^n \in Q$ for some $n \geq 1$. However $a^n \notin Q$ since $a \notin P$. Thus $(b^n)^m = b^{nm} \in Q$ for

some $m \geq 1$ so $b \in P$. Finally P is contained in every prime ideal φ which contains Q. For by Theorem 6-13 $P^n \subseteq Q \subseteq \varphi$ for some n so $P \subseteq \varphi$.

It remains only to show that $A(R/Q) = \{P\}$. Suppose $\varphi \in A(R/Q)$. Then $\varphi \supseteq P$ by the above. We must show $\varphi \subseteq P$. Let $x \in \varphi = (0:A/Q)$ for some prime submodule A/Q of R/Q and $a \in A$, $a \notin Q$. Then $xa \in Q$ implies $x \in \sqrt{Q} = P$ by (2). Thus $\varphi \subseteq P$, that is, $A(R/Q) = \{P\}$.

This shows that the "classical" definition (2) of a primary ideal agrees with our definition. We next recapture the "classical" decomposition theorem.

THEOREM 6-16. (Lasker-Noether Decomposition). Let A be an ideal in a Noetherian ring R. Then $A = \cap_{i=1}^{t} Q_i$ where Q_i is a P_i-primary ideal with $P_i \neq P_j$ for $i \neq j$ and $Q_1 \cap \cdots \cap \hat{Q}_i \cap \cdots \cap Q_t \neq A$ for any $i = 1, \ldots, t$.

Proof. Consider a decomposition of 0 in the module R/A. We have $0 = \cap_{i=1}^{t} Q_i/A$ with $R/A \,/\, Q_i/A \approx R/Q_i$ a P_i-primary module, that is, Q_i is a P_i-primary ideal, $P_i \neq P_j$ for $i \neq j$. Thus $A = \cap_{i=1}^{t} Q_i$ and no subintersection will do.

We next want to investigate the question of whether

the submodules M_i appearing in a decomposition of 0 in M are unique. The answer we will get is that some are and some are not. The technique to be used once again will be localization. We will use the following fact which can be proved in a similar fashion as the corresponding statement for prime ideals: If S is a multiplicative set in R, then there is a one-to-one correspondence between primary ideals in R_S and primary ideals in R which do not meet S.

THEOREM 6-17. Let M be an R-module. Then A(M) is equal to the set of prime ideals P for which there exists an injective R-homomorphism from R/P into M.

Proof. Suppose there exists $0 \to R/P \overset{f}{\to} M$. Then R/P is a prime submodule of M and $(0:R/P) = P$ so $P \in A(M)$.

Now suppose $P \in A(M)$. Then $P = (0:M')$ for some prime submodule $M' \subseteq M$. Let $0 \neq x \in M'$. Then $P = (0:Rx) = \ker \widetilde{x}$ where $\widetilde{x} : R \to M$ is given by $\widetilde{x}(r) = rx$. Thus we get an injective R-homomorphism from R/P into M.

THEOREM 6-18. Let M be an R-module, S a multiplicative set in R, M_S the resulting R_S-module. Then $A(M_S) = \{PR_S \mid P \in A(M) \text{ and } P \cap S = \Phi\}$.

Proof. Let $P \in A(M)$ satisfy $P \cap S = \Phi$. We want to show $PR_S \in A(M_S)$. By Theorem 6-17, it suffices to exhibit an R_S-monomorphism $g : R_S/PR_S \to M_S$. We know there is an R-monomorphism $f : R/P \to M$. Denote by $\bar{r}$ the image of r in R/P. Define $g : R_S/PR_S \to M_S$ by $g(\overline{[r/s]}) = [f(\bar{r})/s]$. It is easily checked that g is well-defined and an R_S-homomorphism. Finally if $g(\overline{[r/s]}) = [f(\bar{r})/s] = 0$, then there exists $s' \in S$ such that $0 = s'f(\bar{r}) = f(\overline{s'r})$, so $s'r \in P$ since ker f = 0. But $s' \notin P$ so $r \in P$. Thus $[r/s] \in PR_S$ so $\overline{[r/s]} = 0$. Therefore g is injective. We conclude that $A(M_S) \supseteq \{PR_S \mid P \in A(M) \text{ and } P \cap S = \Phi\}$.

Set $A(M) = X \cup Y$ where $X = \{P \in A(M) \mid P \cap S = \Phi\}$ and $Y = \{P \in A(M) \mid P \cap S \neq \Phi\}$. Then we can find an exact sequence of R-modules $0 \to M' \xrightarrow{i} M \xrightarrow{j} M'' \to 0$ where $A(M') = X$ and $A(M'') = Y$. We now break up the rest of the proof into a series of steps:

(1) $0 \to M'_S \xrightarrow{\tilde{i}} M_S \xrightarrow{\tilde{j}} M''_S \to 0$ is an exact sequence of R_S-modules. Define $\tilde{i} : M'_S \to M_S$ by $\tilde{i}([m'/s]) = [i(m')/s]$ and $\tilde{j} : M_S \to M''_S$ by $\tilde{j}([m/s]) = [j(m)/s]$. It is easy to check that $\tilde{i}$ and $\tilde{j}$ are R_S-homomorphisms and that $\ker \tilde{i} = 0$, $\operatorname{im} \tilde{j} = M''_S$, and $\ker \tilde{j} = \operatorname{im} \tilde{i}$. (See Exercise 5-5.)

(2) $M''_S = 0$. For $\sqrt{(0:M'')} = \cap_{P \in A(M'')} P = \cap_{P \in Y} P$ so since for each $P \in Y$, $P \cap S \neq \Phi$ and S is a multiplicative set, we get $\sqrt{(0:M'')} \cap S \neq \Phi$. Since for some integer N,

$(\sqrt{(0:M'')})^N \subseteq (0:M'')$ we have that $(0:M'') \cap S \neq \Phi$. Therefore $M''_S = 0$. We conclude from this that $M'_S \approx M_S$.

(3) $\psi : M' \to M'_S$ given by $\psi(m') = [m'/1]$ is injective. For suppose $m' \in \ker \psi$, that is, there exists $s \in S$ such that $sm' = 0$. Then $s \in (0:m') = (0:Rm')$. Let $P \in A(Rm') \subseteq A(M') = X$ since Rm' is a submodule of M'. Then $s \in P$ contradicting $P \cap S = \Phi$.

(4) Let $\varphi \subseteq R_S$ be in $A(M_S) = A(M'_S)$. Then $\varphi = (0:[x/s])$ for some $x \in M'$, $s \in S$. Clearly $\varphi = (0:[x/1])$ and by (3), $\varphi \cap R = (0:x)$ which is a prime ideal in R not meeting S. Call it P. Then $P \in A(M') \subseteq A(M)$. So $\varphi = PR_S \in \{PR_S \mid P \in A(M) \text{ and } P \cap S = \Phi\}$. Therefore $A(M_S) \subseteq \{PR_S \mid P \in A(M) \text{ and } P \cap S = \Phi\}$, completing the proof.

COROLLARY 1. Let M be an R-module and S a multiplicative set in R. Then $M_S = 0$ if and only if $(0:M) \cap S \neq \Phi$.

<u>Proof</u>. $M_S = 0$ if and only if $A(M_S) = \Phi$ if and only if $P \cap S \neq \Phi$ for all $P \in A(M)$ if and only if $S \cap (\cap_{P \in A(M)} P) \neq \Phi$ if and only if $S \cap \sqrt{(0:M)} \neq \Phi$ if and only if $S \cap (0:M) \neq \Phi$.

COROLLARY 2. Every minimal prime containing $(0:M)$ belongs to $A(M)$. In particular, every minimal prime

containing an ideal A of R belongs to A(R/A).

Proof. Let P be a minimal prime containing (0:M). Then $M_P \neq 0$ and so $A(M_P) \neq \Phi$. But the only prime ideal in R_P which contains $(0:M_P) \supseteq (0:M)R_P$ is PR_P. Thus $A(M_P) = \{PR_P\}$ so $P \in A(M)$.

DEFINITION 6-9. A prime $P \in A(M)$ is called an embedded prime if there is a prime $\wp \in A(M)$ such that $P \supset \wp$. Primes in A(M) which are not embedded are called isolated primes.

THEOREM 6-19. (Uniqueness of decomposition of 0 in M) Let $0 = \cap_{i=1}^{t} M_i$ be a primary decomposition of 0 in M. Then the M_i corresponding to isolated primes P_i are unique. In fact $M_i = \ker \varphi_i$ where $\varphi_i : M \to M_{P_i}$.

Proof. Consider the diagram

$$\begin{array}{ccccccccc}
0 & \longrightarrow & M_i & \longrightarrow & M & \xrightarrow{f} & M/M_i & \longrightarrow & 0 \\
 & & \downarrow & & \downarrow \varphi_i & & \downarrow g & & \\
0 & \longrightarrow & (M_i)_{P_i} & \longrightarrow & M_{P_i} & \xrightarrow{h} & (M/M_i)_{P_i} & \longrightarrow & 0
\end{array}$$

where each of the rows is an exact sequence and all the maps involved are the obvious ones. P_i is assumed to be an isolated prime.

(a) $(M_i)_{P_i} = 0$. It suffices to show $A((M_i)_{P_i}) = \Phi$, that is, $\{PR_{P_i} \mid P \in A(M_i) \text{ and } P \cap (R - P_i) = \Phi\} = \Phi$. But $P \in A(M_i)$ implies $P \in A(M)$ which implies $P \cap (R - P_i) \neq \Phi$ since P_i is isolated and $P \neq P_i$ by the nature of M_i.

(b) The map $M/M_i \to (M/M_i)_{P_i}$ is injective. This is proved in exactly the same fashion as step (3) in the previous theorem.

(c) Rewriting the above diagram, we now have

$$\begin{array}{ccccccccc}
 & & & & & & 0 & & \\
 & & & & & & \downarrow & & \\
0 & \longrightarrow & M_i & \longrightarrow & M & \xrightarrow{f} & M/M_i & \longrightarrow & 0 \\
 & & \downarrow & & \downarrow{\varphi_i} & & \downarrow{g} & & \\
 & & 0 & \longrightarrow & M_{P_i} & \xrightarrow{h} & (M/M_i)_{P_i} & \longrightarrow &
\end{array}$$

Therefore

$$\begin{aligned}
M_i &= \ker f = \ker (gf) \text{ since } \ker g = 0 \\
&= \ker (h\varphi_i) \text{ since } h\varphi_i = gf \\
&= \ker \varphi_i \text{ since } \ker h = 0 .
\end{aligned}$$

COROLLARY. Let A be an ideal in a Noetherian ring R

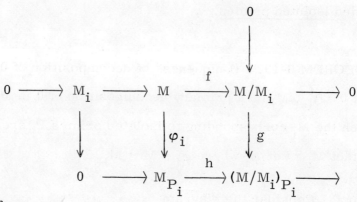

and suppose $A = \cap_{i=1}^{t} Q_i$ is a Lasker-Noether decomposition of A where each Q_i is a P_i-primary ideal. Then the Q_i corresponding to isolated primes P_i are unique. In fact $Q_i = \ker (R \to (R/A)_{P_i})$.

Proof. By Theorem 6-19, $Q_i/A = \ker (R/A \to (R/A)_{P_i})$. Also the preimage of Q_i/A under the canonical map $R \to R/A$ is Q_i.

We conclude this chapter with two more important theorems regarding Noetherian modules.

THEOREM 6-20. (Artin-Rees Theorem) Let R be a commutative Noetherian ring, M a finitely generated R-module, and N an R-submodule of M. Then for any ideal B of R and any positive integer n, there exists a positive integer k(n) such that $B^{k(n)}M \cap N \subseteq B^n N$.

Proof. Since $N/B^nN \subseteq M/B^nN$, we can find an exact sequence $0 \to M'/B^nN \to M/B^nN \to M/M' \to 0$ where $A(M/M') = A(N/B^nN)$ and $A(M'/B^nN) = A(M/B^nN) - A(N/B^nN)$. Thus $A(N \cap M'/B^nN) = A(N/B^nN \cap M'/B^nN)$ $\subseteq A(N/B^nN) \cap A(M'/B^nN) = \Phi$. Hence $N \cap M'/B^nN = 0$ or $N \cap M' \subseteq B^nN$. But $B^nN \subseteq N \cap M'$ so $N \cap M' = B^nN$.

Now

$$B^n \subseteq (0:N/B^nN) \subseteq \cap_{P \in A(N/B^nN)} P$$
$$= \cap_{P \in A(M/M')} P = \sqrt{(0:M/M')}.$$

By Theorem 6-13, there exists an integer t such that $(\sqrt{(0:M/M')})^t \subseteq (0:M/M')$. Setting $k(n) = nt$, we see that $B^{k(n)} \subseteq (0:M/M')$. Thus $B^{k(n)}M \subseteq M'$ so $B^{k(n)}M \cap N \subseteq M' \cap N = B^nN$.

Note. For the reader familiar with the notion of a topological space, the Artin-Rees Theorem has a topological interpretation. Given an R-module M and an ideal B of R, one can supply M with a topology, called the B-adic topology, by taking $\{B^nM \mid n \geq 0\}$ as a system of open sets about 0 in M. If N is a submodule of M, it can be topologized either by $\{B^nN \mid n \geq 0\}$ or by the induced topology from M, that is, $\{B^nM \cap N \mid n \geq 0\}$. The Artin-Rees Theorem states that these topologies are equivalent when R is Noetherian and M finitely generated.

THEOREM 6-21. (Krull Intersection Theorem) Let R be a commutative Noetherian ring and M a finitely generated R-module. Then for any ideal B of R, we have $\cap_{n=1}^{\infty} B^nM = \{x \in M \mid (1-b)x = 0 \text{ for some } b \in B\}$.

Proof. Let $N = \bigcap_{n=1}^{\infty} B^n M$. By the previous theorem $BN \supseteq B^k M \cap N = N \supseteq BN$ for some k. That is, $BN = N$. Let $x \in M$ such that $(1 - b)x = 0$ for some $b \in B$. Then $x = bx = b^2 x = \cdots$ so $x \in \bigcap_{n=1}^{\infty} B^n M = N$. On the other hand, M is a Noetherian module so N is finitely generated, say $N = \langle x_1, \ldots, x_r \rangle$. Since $N = BN$, there exist elements $b_{ij} \in B$, $i = 1, \ldots, r$ and $j = 1, \ldots, r$ such that $x_i = \Sigma_{j=1}^{r} b_{ij} x_j$. Thus $\Sigma_{j=1}^{r} (b_{ij} - \delta_{ij}) x_j = 0$. If we denote by A the r x r matrix (b_{ij}) and by $P_A(t)$ the characteristic polynomial of A, we have $P_A(1)x_j = 0$ for $j = 1, \ldots, r$. But $P_A(t) = t^r + a_1 t^{r-1} + \cdots + a_r$ where $a_1, \ldots, a_r \in B$. Thus $P_A(1) = 1 - b$ for some $b \in B$. Finally if $x \in N$, $x = \Sigma_{j=1}^{r} c_j x_j$ so $P_A(1)x = \Sigma_{j=1}^{r} c_j P_A(1) x_j = 0$ so $(1 - b)x = 0$ for some $b \in B$.

COROLLARY. Let R be a commutative Noetherian local ring with maximal ideal $\mathfrak{m}$. Then $\bigcap_{n=1}^{\infty} \mathfrak{m}^n = 0$.

Proof. Let M = R in Theorem 6-21. Then $\bigcap_{n=1}^{\infty} \mathfrak{m}^n = \{x \in R \mid (1 - b)x = 0 \text{ for some } b \in \mathfrak{m}\}$. But if $b \in \mathfrak{m}$, $1 - b \notin \mathfrak{m}$ and so is a unit. Thus $(1 - b)x = 0$ implies $x = 0$.

Note. Following up on the remark after Theorem 6-20, the import of the above corollary is that a Noetherian local ring

R is a Hausdorff space under the $\mathfrak{m}$-adic topology.

EXERCISES

6-1. Let M be a finitely generated left Λ-module where Λ is a left Noetherian ring. Suppose $\varphi : M \to M$ is a surjective Λ-homomorphism. Prove that φ is injective.

6-2. Let M be a finitely generated module over a Noetherian ring R with A(M) the set of associated primes of M. Prove:

(a) If $P \in A(M)$, then $P = (0:x)$ for some $0 \neq x \in M$.

(b) If $P = (0:x)$, $0 \neq x \in M$, is a prime ideal, then $P \in A(M)$.

6-3. Prove the statement immediately preceding Theorem 6-17 regarding primary ideals in R and R_S.

6-4. Let k be a field and $R = k[X, Y, Z]/(Z^2 - XY)$. Denote by x, y, z the canonical images of X, Y, Z in R. Prove:

(a) $B = (x, y, z)$ is a maximal ideal of R and B^2 is a B-primary ideal.

(b) $P = (x, z)$ is a prime ideal of R and $A = (x)$ is a P-primary ideal.

(c) $P^2 = A \cap B^2$ is a primary decomposition of P^2 in R.

Thus we have an example of a power of a prime ideal which is not primary.

6-5. Let R be a Noetherian ring and P a maximal ideal of R. Prove that for any $n \geq 1$, P^n is a P-primary ideal.

6-6. Let R be a Noetherian ring and A the intersection of all minimal prime ideals of R. Prove that $A^n = 0$ for some positive integer n and that every element of $1 - A = \{1 - a \mid a \in A\}$ is a unit in R.

6-7. Let R be a Noetherian ring and $(0) = Q_1 \cap \cdots \cap Q_n$ a primary decomposition of the zero ideal in R where each Q_i is P_i-primary. Prove that the set of zero divisors of R equals the set-theoretic union of $P_1, \ldots, P_n$.

6-8. Let R be a Noetherian ring, P a nonmaximal prime ideal of R, $Q \neq P$ a P-primary ideal. Prove that there exists an ideal A with $P \supset A \supset Q$ which is not a primary ideal.

6-9. Let R be a Noetherian domain and P a nonzero prime of R. Prove that $P^2 \neq P$.

6-10. Let Λ be a regular left Noetherian ring. Prove that every left ideal of Λ is principal. (Hint: It suffices to show that every left ideal generated by two elements is principal. Use Exercise 4-10.)

CHAPTER 7

DEDEKIND DOMAINS

In view of the introductory remarks to Chapter 6 and the results obtained in that chapter, we are still left with the question of describing a class of rings in which every ideal can be written (preferably uniquely) as a product of prime ideals. Since Noetherian rings yield a partial result in this direction, by replacing "product" with "intersection" and "prime" with "primary", we might expect a subclass of Noetherian rings to be the sought after type. In this chapter we characterize these rings called Dedekind domains.

DEFINITION 7-1. Let R be an integral domain with quotient field K. An R-module $0 \neq B \subseteq K$ is called a <u>fractionary ideal</u> if there exists $d \neq 0$ in R such that $B \subseteq d^{-1} R$.

Note. 1. If B is a fractionary ideal, then $B = d^{-1}A$ where A is an ordinary ideal of R. Namely $A = \{x \in R \mid d^{-1}x \in B\}$.

2. Every ordinary ideal $0 \neq A \subseteq R$ is a fractionary ideal by taking $d = 1$. These will now be called integral ideals.

3. The addition, multiplication, intersection, and quotient of fractionary ideals can be defined as they were for integral ideals in Chapter 2. The relationships among these operations carry over to fractionary ideals.

4. The ideal R acts as a unit in the multiplication of fractionary ideals.

5. If M is an R-module and B is a fractionary ideal of R, then $BM = \{\Sigma_{\text{finite}}\, b_i m_i \mid b_i \in B,\ m_i \in M\}$. Since $B \subseteq K = R_0$ each summand $b_i m_i$ is an element of M_0 and the addition should be interpreted as taking place in M_0. Thus BM is a certain R-submodule of M_0.

DEFINITION 7-2. A fractionary ideal A is called invertible if there exists a fractionary ideal A^{-1} such that $AA^{-1} = R$.

THEOREM 7-1. If a fractionary ideal A is invertible, then $A^{-1} = (R:A) = \{x \in K \mid xA \subseteq R\}$. In particular, A^{-1} is unique.

Proof. Suppose $AA' = R$. Then $A' \subseteq (R:A)$. On the other hand, $(R:A) = (R:A)R = (R:A)AA' \subseteq RA' = A'$ so $A' = (R:A)$.

THEOREM 7-2. If every integral ideal of R is invertible, then every fractionary ideal is invertible.

Proof. Let B be a fractionary ideal. Then $B = d^{-1}A$, where A is an integral ideal. If A^{-1} is the inverse of A, then dA^{-1} is the inverse of B so B is invertible.

THEOREM 7-3. Let R be an integral domain in which every ideal is a unique product of prime ideals and P an invertible integral prime ideal of R. Then P is maximal.

Proof. Let $a \in R - P$ and set $B = (P, a)$, $C = (P, a^2)$ and $D = (P^2, a)$. We want to show $B = R$. Since P is invertible, this is equivalent to showing $PB = P$. Let $B = \Pi_{i=1}^{s} P_i$ and $C = \Pi_{j=1}^{t} Q_j$ be the prime factorizations of B and C. Set $\overline{R} = R/P$ and $\overline{I} = I/P$ for any ideal I of R. Then

$$\prod_{i=1}^{s} \overline{P}_i^{\,2} = \overline{B}^2 = \overline{C} = \prod_{j=1}^{t} \overline{Q}_j .$$

By unique factorization in $\overline{R}$, $t = 2s$ and by relabeling, we can assume that $\overline{Q}_{2i} = \overline{Q}_{2i-1} = \overline{P}_i$ for $i = 1, \ldots, s$. Thus

$Q_{2i} = Q_{2i-1} = P_i$ for $i = 1, \ldots, s$ so $B^2 = C$. Therefore $P \subseteq C = B^2 \subseteq D$. If $x \in P$, $x = y + ra$ for some $y \in P^2$, $r \in R$. Then $ra = x - y \in P$ and $a \notin P$ so $r \in P$. Therefore $P \subseteq (P^2, Pa) = PB$. The inclusion $PB \subseteq P$ is obvious and so $PB = P$.

THEOREM 7-4. Let R be a local principal ideal domain with maximal ideal $\mathfrak{m} = (\pi)$. Then every nonzero ideal in R is a power of $\mathfrak{m}$ and so $\mathfrak{m}$ is the only prime ideal of R.

Proof. Let $0 \neq A = (\alpha)$ be an ideal of R. Then $A \subseteq \mathfrak{m}$ which implies π divides α. Since R is a unique factorization domain, there exists an integer $n \geq 1$ such that π^n divides α but π^{n+1} does not divide α. Thus $\alpha = u\pi^n$ where $u \notin (\pi)$. Thus $u \notin \mathfrak{m}$ and so u is a unit in R. Thus $A = (\alpha) = (u\pi^n) = (\pi^n) = \mathfrak{m}^n$. A can be a prime ideal only if $n = 1$, that is, $A = \mathfrak{m}$.

COROLLARY. Let R be a commutative ring such that R_P is a principal ideal domain for every prime ideal P of R. Then every nonzero prime in R is maximal.

Proof. Let $0 \neq P$ be prime in R and $M \supseteq P$, maximal. Then R_M is a local principal ideal domain with nonzero prime ideals PR_M and MR_M. By Theorem 7-4, $PR_M = MR_M$

so P = M. Therefore P is maximal.

THEOREM 7-5. Let R be a Noetherian local ring and P a finitely generated projective R-module. Then P is a free R-module.

Proof. Let $\{x_1, \ldots, x_n\}$ be a minimal generating set for P, F a free R-module of rank n. Thus we have an exact sequence $0 \to X \to F \to P \to 0$ where $F = P \oplus X$ and X is also a finitely generated projective R-module since R is Noetherian. Then $F/\mathfrak{m}F = P/\mathfrak{m}P \oplus X/\mathfrak{m}X$ where $\mathfrak{m}$ is the maximal ideal of R. Since $R/\mathfrak{m}$ is a field, each of these modules is a free $R/\mathfrak{m}$-module and we know that the rank of $F/\mathfrak{m}F$ is n. We will show that the rank of $P/\mathfrak{m}P$ is also n from which it follows that $X/\mathfrak{m}X = 0$. This is done by showing that $\{\bar{x}_1, \ldots, \bar{x}_n\}$ is an $R/\mathfrak{m}$-basis for $P/\mathfrak{m}P$.

Certainly $\bar{x}_1, \ldots, \bar{x}_n$ span $P/\mathfrak{m}P$ over $R/\mathfrak{m}$. Suppose $\Sigma_{i=1}^n \bar{r}_i\bar{x}_i = 0$ for some $\bar{r}_i \in R/\mathfrak{m}$. If not all $\bar{r}_i = 0$, say $\bar{r}_1 \neq 0$ (relabel if necessary). That is, $r_1 \notin \mathfrak{m}$. Then $\Sigma_{i=1}^n r_ix_i = \Sigma_{i=1}^n m_ix_i$ for some $m_i \in \mathfrak{m}$ or $\Sigma_{i=1}^n (r_i - m_i)x_i = 0$. But $r_1 - m_1 \notin \mathfrak{m}$ since $r_1 \notin \mathfrak{m}$ and so is a unit in R. Multiplying by the inverse of $r_1 - m_1$ we see that $x_1 \in \langle x_2, \ldots, x_n \rangle$, contradicting the minimality of the generating set $\{x_1, \ldots, x_n\}$.

The final step of the proof is to show that $X/\mathfrak{m}X = 0$ implies $X = 0$. Let $\{y_1, \ldots, y_t\}$ be a minimal generating set of X. Then $y_1 = \Sigma_{j=1}^{t} m_j y_j$ for some $m_j \in \mathfrak{m}$. So $(1 - m_1)y_1 = \Sigma_{j=2}^{t} m_j y_j$ where $1 - m_1$ is a unit in R. Thus $y_1 \in \langle y_2, \ldots, y_t \rangle$, contradiction. Thus a minimal generating set for X is empty, that is, $X = 0$. Therefore $F = P$ is a free R-module.

Note. The final step of this proof follows immediately from Nakayama's lemma which is proven in the next chapter. The proof given there is virtually identical to the argument here.

THEOREM 7-6. Let R be an integral domain. The following statements are equivalent:

(1) Every ideal in R is a unique product of prime ideals.

(2) Every fractionary ideal of R is invertible.

(3) R is Noetherian and every ideal of R is R-projective.

(4) R is Noetherian and R_P is a principal ideal domain for every prime ideal P of R.

Proof. (1) implies (2). By Theorem 7-2, it suffices to show that every integral ideal is invertible. Since, by hypothesis, every integral ideal is a product of prime ideals, it is enough to show that every integral prime ideal is invertible. Let P be an integral prime ideal and $x \in P$. Then $Rx = \Pi_{i=1}^{n} P_i$ for prime ideals P_i. For each $j = 1, \ldots, n$, we have $R = (Rx^{-1})Rx = P_j(Rx^{-1}) \Pi_{\substack{i=1 \\ i \neq j}}^{n} P_i$ so P_j is invertible. By Theorem 7-3, each P_j is maximal. But $P \supseteq \Pi_{i=1}^{n} P_i$ so $P \supseteq P_j$ for some j, whence $P = P_j$. Thus P is invertible.

(2) implies (3). Let A be an ideal of R with A^{-1} its inverse. Then $AA^{-1} = R$ implies that there exist elements $x_1, \ldots, x_n \in A$, $y_1, \ldots, y_n \in A^{-1}$ such that $1 = \Sigma_{i=1}^{n} x_i y_i$. Then if $x \in A$, $x = x \cdot 1 = \Sigma_{i=1}^{n} (xy_i)x_i \in (x_1, \ldots, x_n)$ so A is finitely generated. Therefore R is Noetherian.

We wish to show A is a projective R-module. Consider the R-homomorphism $\psi : R^n \to A$ given by $\psi(r_1, \ldots, r_n) = \Sigma_{i=1}^{n} r_i x_i$. Define an R-homomorphism $\varphi : A \to R^n$ by $\varphi(x) = (y_1 x, \ldots, y_n x)$. Then $(\psi \circ \varphi)(x) = \psi(y_1 x, \ldots, y_n x) = \Sigma_{i=1}^{n} y_i x x_i = x$ so $\psi \circ \varphi$ is the identity map on A. Hence φ is a cross-section for the exact sequence $0 \to \ker \psi \to R^n \overset{\psi}{\to} A \to 0$. By Theorem 5-3, A is a direct summand of R^n and so by Theorem 5-4 is projective.

(3) implies (4). Suppose R is Noetherian and every ideal is projective. Let P be a prime ideal of R. If $P = 0$, then R_P is the quotient field of R which is a principal ideal domain. If $P \neq 0$, let A' be an ideal of R_P. Then

$A' = AR_P$ where $A = A' \cap R$. Since A is a projective R-module and localization preserves exact sequences, A' is a projective R_P-module. But a projective module over a local ring is free so A' is a free R_P-module, say $A' = R_Px_1 \oplus R_Px_2 \oplus \cdots \oplus R_Px_t$. Then $x_1x_2 = x_2x_1 \in R_Px_1 \cap R_Px_2 = 0$ so $x_1 = 0$ or $x_2 = 0$. That is, A' is a principal ideal.

(4) implies (1). Let A be an ideal of R with $A = Q_1 \cap \cdots \cap Q_t$ where Q_i is P_i-primary. Since each P_i is maximal, they are all isolated primes so the Q_i are unique. We wish to show that $Q_i = P_i^{\nu_i}$ for some $\nu_i \geq 1$. Dropping the subscript, let Q be one of the Q_i. Then QR_P is a PR_P-primary ideal in R_P. By Theorem 7-4, $QR_P = (PR_P)^\nu = P^\nu R_P$. Since P^ν is a P-primary ideal in R (see Exercise 6-5), we have $Q = QR_P \cap R = P^\nu R_P \cap R = P^\nu$. Thus $A = P_1^{\nu_1} \cap \cdots \cap P_t^{\nu_t}$. However since the P_i are distinct maximal ideals, they are comaximal. Hence so are $P_1^{\nu_1}, \ldots, P_t^{\nu_t}$. Thus products and intersections coincide, that is, $A = P_1^{\nu_1} \cdots P_t^{\nu_t}$.

DEFINITION 7-3. A ring satisfying one (hence all) of the above is called a <u>Dedekind domain.</u>

COROLLARY. Every nonzero prime ideal in a Dedekind

domain is maximal.

EXERCISES

7-1. Let R be a Dedekind domain, M an R-module, and (0:M) a prime ideal of R. Prove that M is a prime module.

7-2. Let S be a multiplicative set in a Dedekind domain R. Provc that R_S is a Dedekind domain.

7-3. Let R be a Dedekind domain and a unique factorization domain. Prove that R is a principal ideal domain. (Hint: It suffices to show that each prime ideal is principal.)

7-4. Let R be a Dedekind domain with a finite number of prime ideals. Prove that R is a principal ideal domain. (Hint: Use Exercise 6-9 to find $a \in P$, $a \notin P^2$. Then use the Chinese Remainder Theorem to find $x \in P$, $x \notin P^2$, $x \notin Q$ for any prime $Q \neq P$. Finally show that $P = (x)$.)

7-5. Let A be a nonzero ideal of a Dedekind domain R. Prove that R/A is a principal ideal domain.

7-6. Prove that $\mathbb{Z}[\sqrt{-5}]$ is a Dedekind domain.

7-7. Let R be an integral domain with quotient field K. An element $\alpha \in K$ is called <u>integral</u> over R if there exists a monic polynomial $f(X) \in R[X]$ such that $f(\alpha) = 0$. Prove that $\alpha \in K$ is integral over R if and only if $R[\alpha]$ is a finitely generated R-module.

7-8. An integral domain R is called <u>integrally closed</u> if every $\alpha \in K$ which is integral over R is in R. Prove that every Dedekind domain is integrally closed.

7-9. Let K be a field and $K^* = K - \{0\}$. A map $v : K \to \mathbb{Z} \cup \{\infty\}$ is called a <u>discrete valuation</u> if

(1) $v(0) = \infty$;

(2) $v : K^* \to \mathbb{Z}$ is a group epimorphism;

(3) $v(x+y) \geq \min(v(x), v(y))$.

If v is a discrete valuation of K, then $R_v = \{x \in K \mid v(x) \geq 0\}$ is called the <u>discrete valuation ring</u> of v. Suppose R is a Dedekind domain with quotient field K and P a nonzero prime ideal of R. Prove that there exists a discrete valuation v_P of K with $R_{v_P} = R_P$.

7-10. Using the notation of the previous exercise, prove that $PR_P = \{x \in R_{v_P} \mid v_P(x) > 0\}$.

CHAPTER 8

ARTIN RINGS AND MODULES

In this chapter and the following chapter Λ will denote an arbitrary ring with unit element, not necessarily commutative.

LEMMA. Let A be a left ideal of a ring Λ such that every element of the set $1 + A = \{1 + a \mid a \in A\}$ has a left inverse in Λ. Then every element of $1 + A$ is a unit in Λ.

Proof. Let $a \in A$ and $b \in \Lambda$ a left inverse of $1 + a$, that is, $b(1 + a) = 1$. Thus $b = 1 - ba \in 1 + A$ so there exists $b' \in \Lambda$ such that $b'b = 1$. Then $1 + a = b'b(1 + a) = b'$ so $(1 + a)b = b'b = 1$. Thus b is also a right inverse for $1 + a$ so $1 + a$ is a unit.

DEFINITION 8-1. A nonzero left Λ-module E is called

simple if the only left Λ-submodules of E are 0 and E.

THEOREM 8-1. The following ideals in a ring Λ are identical:

(1) $R_1 = \cap_E (0:E)$ where E ranges over all simple left Λ-modules.

(2) $R_2 = \cap_{\mathfrak{m}} \mathfrak{m}$ where $\mathfrak{m}$ ranges over all maximal left ideals of Λ.

(3) $R_3 = \cup_A A$ where A ranges over all left ideals A such that $1 + A$ consists solely of units.

(4) $R_4 = \{x \in \Lambda \mid 1 + \lambda x \mu$ is a unit in Λ for all $\lambda, \mu \in \Lambda\}$.

(5) $R_5 = \cap_E (0:E)$ where E ranges over all simple right Λ-modules.

(6) $R_6 = \cap_{\mathfrak{m}} \mathfrak{m}$ where $\mathfrak{m}$ ranges over all maximal right ideals of Λ.

(7) $R_7 = \cup_A A$ where A ranges over all right ideals A such that $1 + A$ consists solely of units.

Proof. We will show $R_1 = R_2 = R_3 = R_4$ which by symmetry is sufficient.

(1) $R_1 = R_2$. First observe that E is a simple left Λ-module if and only if $E \approx \Lambda/\mathfrak{m}$ for some maximal left ideal $\mathfrak{m}$. For let $0 \neq \alpha \in E$. Then $0 \neq \Lambda\alpha \subseteq E$ implies $E = \Lambda\alpha$. Consider the Λ-epimorphism $\tilde{\alpha} : \Lambda \to E$ given

$\widetilde{\alpha}(\lambda) = \lambda\alpha$. Then $E \approx \Lambda/\ker\widetilde{\alpha}$. Since E is simple, $\ker\widetilde{\alpha}$ must be a maximal left ideal of Λ. The converse is immediate. Thus

$$R_1 = \cap_{E:\text{simple}}(0:E) = \cap_{\mathfrak{m}:\text{maximal}}(0:\Lambda/\mathfrak{m})$$
$$\subseteq \cap_{\mathfrak{m}:\text{maximal}}\mathfrak{m} = R_2.$$

To prove the opposite inclusion note that $(0:E) = \cap_{\alpha\in E}(0:\alpha)$ and that each $(0:\alpha)$ is a maximal left ideal of Λ. For suppose $x \notin (0:\alpha)$. Then $0 \neq (\Lambda x)\alpha \subseteq E$ so $(\Lambda x)\alpha = E$ since E is simple. In particular $\alpha = \lambda x\alpha$ for some $\lambda \in \Lambda$, that is, $(1 - \lambda x)\alpha = 0$. Thus $1 - \lambda x \in (0:\alpha)$ so $(0:\alpha) + \Lambda x = \Lambda$. That is, $(0:\alpha)$ is a maximal left ideal. Hence we have

$$R_1 = \cap_{E:\text{simple}}(0:E) = \cap_{E:\text{simple}}\cap_{\alpha\in E}(0:\alpha)$$
$$\supseteq \cap_{\mathfrak{m}:\text{maximal}}\mathfrak{m} = R_2.$$

(2) $R_2 = R_3$. Let $x \in R_2$ and consider the left ideal $\Lambda(1+x) = A$. If A is a proper left ideal of Λ, then $A \subseteq \mathfrak{m}$ for some maximal left ideal of Λ by Zorn's lemma. But then $1 + x \in A \subseteq \mathfrak{m}$ and $x \in R_2 \subseteq \mathfrak{m}$ implies $1 \in \mathfrak{m}$ which is a contradiction. Thus $\Lambda(1+x) = \Lambda$. That is, every element of $1 + R_2$ has a left inverse in Λ, hence by the lemma is a unit in Λ. Thus $R_2 \subseteq R_3$. On the other hand, suppose $x \in R_3$ and $\mathfrak{m}$ is a maximal left ideal of Λ

such that $x \notin \mathfrak{m}$. Then $(\mathfrak{m}, x) = \Lambda$, that is, $1 = \lambda x + m$ for some $\lambda \in \Lambda$, $m \in \mathfrak{m}$. Then $m = 1 - \lambda x$ is a unit in Λ, that is, $\mathfrak{m} = \Lambda$. Contradiction. Hence $x \in \mathfrak{m}$ so $x \in R_2$. Therefore $R_3 \subseteq R_2$.

(3) $R_3 = R_4$. Clearly $R_4 \subseteq R_3$ by setting $\mu = 1$ in the definition of R_4. To show $R_3 \subseteq R_4$, observe that since $R_3 = R_1$ and R_1 is a two-sided ideal, R_3 is also two-sided. Hence if $x \in R_3$, $\lambda x \mu \in R_3$ so $1 + \lambda x \mu$ is a unit, that is, $x \in R_4$.

DEFINITION 8-2. The ideal described by any of the equivalent formulations in Theorem 8-1 is called the <u>radical of Λ</u> and is denoted by $R(\Lambda)$.

THEOREM 8-2. (Nakayama's lemma) Let M be a finitely generated left Λ-module. Then $M = 0$ if and only if $M/R(\Lambda)M = 0$.

<u>Proof</u>. Certainly $M = 0$ implies $M/R(\Lambda)M = 0$. On the other hand, let $x_1, \ldots, x_n$ be a minimal generating set of M where $M \neq 0$ implies $n \geq 1$. Then $M/R(\Lambda)M = 0$ says we can express $x_1 = \Sigma_{i=1}^{n} r_i x_i$ where $r_i \in R(\Lambda)$. Thus $(1 - r_1)x_1 = \Sigma_{i=2}^{n} r_i x_i$. But $r_1 \in R(\Lambda)$ implies $1 - r_1$ is a unit in Λ. Multiplying the equation on the left by the

inverse of $1 - r_1$ gives $x_1 \in \langle x_2, \ldots, x_n \rangle$, contradicting the choice of $\{x_1, \ldots, x_n\}$. Thus $n = 0$, that is, $M = 0$.

DEFINITION 8-3. A left ideal A in a ring Λ is said to be <u>nilpotent</u> if $A^n = 0$ for some integer $n \geq 1$. A left ideal A is called a <u>nil ideal</u> if for each $x \in A$, there exists an integer $n(x) \geq 1$ such that $x^{n(x)} = 0$.

<u>Note</u>. Clearly every nilpotent ideal is nil. For if $A^n = 0$, one can choose $n(x) = n$ for every $x \in A$. The converse however is false. Consider the ring $k[X_1, X_2, X_3, \ldots]$ of polynomials in an infinite number of variables. Let B be the ideal generated by $\{X_i^{i+1} \mid i = 1, 2, 3, \ldots\}$ and set $R = k[X_1, X_2, \ldots]/B$. Denote by x_i the image of X_i in R and consider the ideal $A = (x_1, x_2, \ldots)$ in R. A is nil since for each i, $x_i^{i+1} = 0$. However A is not nilpotent. For given any $n > 0$, $0 \neq x_n^n \in A^n$.

THEOREM 8-3. If A is a nil left ideal of Λ, then $A \subseteq R(\Lambda)$.

<u>Proof.</u> It suffices to show that every element of $1 + A$ has a left inverse in Λ. Suppose $a \in A$ and $a^n = 0$. Then

$$(1 - a + a^2 - a^3 + \cdots + (-1)^{n-1} a^{n-1})(1 + a) = 1 + (-1)^{n-1} a^n = 1$$

and so $1 + a$ has a left inverse.

COROLLARY. If A is a nilpotent left ideal of Λ, then $A \subseteq R(\Lambda)$.

THEOREM 8-4. Let Λ be a ring and M a left Λ-module. The following conditions are equivalent:

(1) Every strictly descending chain of Λ-submodules of M, $M_1 \supset M_2 \supset M_3 \supset \cdots$ is finite (the descending chain condition).

(2) Every nonempty collection of left Λ-submodules of M has a minimal element (the minimum condition).

Proof. The proof is the same as that of Theorem 4-2, except that all inclusions are reversed.

DEFINITION 8-4. A left Λ-module M satisfying one (hence both) of the above conditions is called left Artin. A ring Λ is called left Artin if it is left Artin as a left Λ-module. Right Artin is defined in a similar way for modules and rings.

THEOREM 8-5. Let Λ be a left Artin ring. Then $R(\Lambda)$ is a nilpotent ideal.

Proof. Let $R = R(\Lambda)$ and consider the descending chain $R \supseteq R^2 \supseteq R^3 \supseteq \cdots$. Then for some n, $R^i = R^n$ for all

$i \geq n$. Suppose that $R^n \neq 0$ and let $\mathcal{S} = \{0 \neq A \subseteq R^n \mid A$ is a left ideal of Λ and $R^n A \neq 0\}$. We know $\mathcal{S}$ is nonempty since $R^n \in \mathcal{S}$. Let A be a minimal element of $\mathcal{S}$. Then there exists $a \in A$ such that $B = R^n a \neq 0$. Then $R^n B = R^{2n} a = R^n a \neq 0$ and $B \subseteq A$ so by the choice of A, we know $B = A$. Thus $\Lambda a \subseteq A = R^n a \subseteq \Lambda a$ so $A = \Lambda a$, that is, A is a principal ideal. In particular A is a finitely generated left Λ-module and $A = R^n a = R^{n+1} a = RA$. So by Nakayama's lemma, $A = 0$. Contradiction. Thus $R^n = 0$, that is, $R(\Lambda)$ is nilpotent.

DEFINITION 8-5. A nonzero left ideal I in a ring Λ is called <u>simple</u> if it is simple as a left Λ-module. A simple left ideal is sometimes called a minimal left ideal since it contains no subideals other than 0 and itself.

THEOREM 8-6. Let I be a simple left ideal of Λ. Then I is a direct summand of Λ if and only if $I^2 \neq 0$.

<u>Proof</u>. (1) Suppose $\Lambda = I \oplus P$ where P is some left Λ-module. Let $\pi : \Lambda \to I$ be the projection. Then $0 \neq \pi(1) = \pi(\pi(1)) = \pi(\pi(1) \cdot 1) = \pi(1)\pi(1) = (\pi(1))^2 \in I^2$ so $I^2 \neq 0$.

(2) Suppose $I^2 \neq 0$. Then there exists $x \in I$ such

that $Ix \neq 0$, whence $Ix = I$. Thus $yx = x$ for some element $y \in I$. Then $y^2 = y$. For $y^2x = y(yx) = yx$ implies $(y^2 - y)x = 0$. Then if $y^2 - y \neq 0$, $0 \neq \Lambda(y^2 - y) \subseteq I$ gives $\Lambda(y^2 - y) = I$. But this implies $0 = \Lambda(y^2 - y)x = Ix$, contradicting the choice of x. Furthermore $0 \neq \Lambda y \subseteq I$ implies $\Lambda y = I$.

Now consider the exact sequence $0 \to I \xrightarrow{i} \Lambda \to \Lambda/I \to 0$ where i is the inclusion map and define a Λ-homomorphism $\tilde{y} : \Lambda \to I$ given by $\tilde{y}(\lambda) = \lambda y$. Then $\tilde{y} \circ i$ is the identity map on I. For if $x \in I$, $x = \lambda y$ for some $\lambda \in \Lambda$. Then $(\tilde{y} \circ i)(x) = \tilde{y}(\lambda y) = \lambda y^2 = \lambda y = x$. Thus the sequence splits and $\Lambda \approx I \oplus \Lambda/I$.

THEOREM 8-7. Let Λ be a left Artin ring and suppose $R(\Lambda) = 0$. Then Λ is a finite direct sum of simple left ideals.

Note. An Artin ring whose radical is zero is called semisimple. We will have a great deal more to say about such rings in the next chapter.

Proof. Let $\mathcal{S}$ be the collection of nonzero left ideals A of Λ which cannot be written as a finite direct sum of simple left ideals of Λ. If $\mathcal{S}$ is empty, we are done. For then the left ideal Λ is a finite direct sum of simple left ideals.

If $\mathcal{S}$ is nonempty, then we can choose a minimal element A of $\mathcal{S}$ since Λ is left Artin. Now let $\mathfrak{J}$ be the collection of nonzero left ideals of Λ contained in A. Then $\mathfrak{J}$ is nonempty so again we can choose a minimal element I of $\mathfrak{J}$. I is a simple left ideal of Λ and $I^2 \neq 0$. For if $I^2 = 0$, then I is nilpotent so by the corollary to Theorem 8-3, $I \subseteq R(\Lambda) = 0$, that is, $I = 0$, which is a contradiction. Thus by Theorem 8-6, I is a direct summand of Λ, say $\Lambda = I \oplus X$. Let $\pi : \Lambda \to I$ be the natural projection and consider the exact sequence of left Λ-modules $0 \to I \overset{i}{\to} A \to A/I \to 0$ where i is the inclusion map. Define $f : A \to I$ by $f(a) = \pi(a)$. Then if $x \in I$, $(f \circ i)(x) = f(x) = \pi(x) = x$ so f is a retraction. Thus the sequence splits so $A \approx I \oplus A/I$. We can find a left Λ-submodule A' of A, that is, a left ideal of Λ, isomorphic to A/I such that $A = I \oplus A'$. Note that $A' \neq 0$. Otherwise $A = I$, a finite direct sum of simple left ideals. Also $A' \subset A$ since $I \neq 0$. By the minimality of A, A' is a finite direct sum of simple left ideals. Hence so is $A = I \oplus A'$. Contradiction. Thus $\mathcal{S}$ must have been empty.

THEOREM 8-8. Let M be a left Λ-module where Λ is a finite direct sum of simple left ideals. Then

(1) Every left Λ-submodule of M is a direct summand of M.

(2) M is a direct sum of simple left Λ-submodules.

(3) If M is left Artin, then M is a finite direct sum of simple left Λ-submodules.

Proof. (1) Suppose $\Lambda = \bigoplus_{j=1}^{n} I_j$ where each I_j is a simple left ideal. Let M' be a left Λ-submodule of M. Let $\mathcal{S} = \{N \subseteq M \mid N$ is a left Λ-submodule of M and $N \cap M' = 0\}$. $\mathcal{S}$ is nonempty since $0 \in \mathcal{S}$ and is inductive. Let M'' be a maximal element of $\mathcal{S}$ by Zorn's lemma and consider $M' \oplus M'' \subseteq M$. If $x \in M$, $x \notin M' \oplus M''$, then there exists j, $1 \le j \le n$, such that $I_j x \not\subseteq M' \oplus M''$. Otherwise $x \in \Lambda x \subseteq M' \oplus M''$ contrary to the choice of x. Consider the exact sequence $0 \to \ker \tilde{x} \to I_j \xrightarrow{\tilde{x}} I_j x \to 0$ where $\tilde{x}(a) = ax$. Since $I_j x \ne 0$ and I_j is simple, $\ker \tilde{x} = 0$. Thus $I_j x \approx I_j$ and so is also a simple left Λ-module. Therefore $I_j x \cap (M' \oplus M'') = 0$ which implies that $(I_j x \oplus M'') \cap M' = 0$, contradicting the choice of M''. We conclude that $M = M' \oplus M''$ so M' is a direct summand of M.

(2) We begin by showing that M contains a simple left Λ-submodule. Let $0 \ne x \in M$ and set $\mathcal{S} = \{M' \mid M'$ is a left Λ-submodule of M and $x \notin M'\}$. Then $\mathcal{S}$ is nonempty since $0 \in \mathcal{S}$ and is inductive. So by Zorn's lemma, let M_1 be a maximal element of $\mathcal{S}$. By (1), we have $M = M_1 \oplus M_2$ and the claim is that M_2 is a simple left

Λ-module. For suppose $0 \neq M_2' \subseteq M_2$. Then again by (1), $M_2 = M_2' \oplus M_2''$. Then $x \notin M_1 = (M_1 \oplus M_2') \cap (M_1 \oplus M_2'')$ so either $x \notin M_1 \oplus M_2' \supset M_1$ or $x \notin M_1 \oplus M_2'' \supset M_1$, contradicting the choice of M_1. Therefore M_2 is a simple left Λ-submodule of M.

Now let $\mathfrak{J}$ be the collection of all index sets A such that for each $\alpha \in A$ we have a simple left Λ-submodule M_α of M such that $\bigoplus_{\alpha \in A} M_\alpha \subseteq M$. We know that $\mathfrak{J}$ is nonempty, that is, it contains the index set $\{1\}$. Also $\mathfrak{J}$ is inductive. By Zorn's lemma, let Δ be a maximal element of $\mathfrak{J}$ and set $M' = \bigoplus_{\alpha \in \Delta} M_\alpha \subseteq M$. If $M' \neq M$, then by (1) $M = M' \oplus M''$ where $M'' \neq 0$. Then let M_β be a simple left Λ-submodule of M'' (hence also of M) and set $\Delta' = \Delta \cup \{\beta\}$. Then $\bigoplus_{\gamma \in \Delta'} M_\gamma \subseteq M$ where all M_γ are simple, contradicting the choice of Δ. Therefore $M' = M$.

(3) Suppose $M = \bigoplus_{\alpha \in A} M_\alpha$ where the M_α are simple and A is infinite. Then $M \supseteq \bigoplus_{i=1}^{\infty} M_i \supset \bigoplus_{i=2}^{\infty} M_i \supset \cdots$ is an infinite strictly descending chain of left Λ-submodules of M where the M_i are a countable subset of the M_α. This is a contradiction if M is left Artin.

DEFINITION 8-6. Let M be a left Λ-module. A finite descending chain of left Λ-submodules $M = M_0 \supseteq M_1 \supseteq M_2 \supseteq \cdots \supseteq M_r = 0$ is called a normal series for M of length r. The factor modules M_i/M_{i+1}, $i = 0, \ldots, r-1$ are called the factors of the normal series.

DEFINITION 8-7. Two normal series for a left Λ-module M are called equivalent if there is a bijection between the sets of factors in such a way that corresponding factors are Λ-isomorphic.

DEFINITION 8-8. A Jordan-Hölder series for M is a normal series $M = M_0 \supset M_1 \supset M_2 \supset \cdots \supset M_r = 0$ in which each factor is a simple Λ-module. In other words, each M_i is a maximal Λ-submodule of M_{i-1}, $i = 1, \ldots, r$.

DEFINITION 8-9. A normal series $M = M_0 \supseteq \cdots \supseteq M_r = 0$ is said to be a refinement of a normal series $M = N_0 \supseteq \cdots \supseteq N_s = 0$ if every submodule appearing in the second series appears in the first as well.

THEOREM 8-9. (Schreier Refinement Theorem) Any two normal series for a left Λ-module M have equivalent refinements.

Proof. Let $M = M_0 \supseteq \cdots \supseteq M_r = 0$ and $M = N_0 \supseteq \cdots \supseteq N_s = 0$

be two normal series for M. Set $M_{i,j} = M_{i+1} + (M_i \cap N_j)$ for $i = 0, \ldots, r-1$, $j = 0, \ldots, s$ and $N_{j,i} = N_{j+1} + (M_i \cap N_j)$ for $i = 0, \ldots, r$, $j = 0, \ldots, s-1$. Thus we have $M_{i,s} = M_{i+1} = M_{i+1,0}$ and $N_{j,r} = N_{j+1} = N_{j+1,0}$ yielding the following two normal series for M:

$$\begin{aligned} M = M_{0,0} &\supseteq M_{0,1} \supseteq \cdots \supseteq M_{0,s} \\ &= M_{1,0} \supseteq M_{1,1} \supseteq \cdots \supseteq M_{1,s} \\ &= M_{2,0} \supseteq \cdots \supseteq M_{r-1,s-1} \supseteq 0; \end{aligned}$$

$$\begin{aligned} M = N_{0,0} &\supseteq N_{0,1} \supseteq \cdots \supseteq N_{0,r} \\ &= N_{1,0} \supseteq N_{1,1} \supseteq \cdots \supseteq N_{1,r} \\ &= N_{2,0} \supseteq \cdots \supseteq N_{s-1,r-1} \supseteq 0. \end{aligned}$$

However by Theorem 5-6, $M_{i,j}/M_{i,j+1} \approx N_{j,i}/N_{j,i+1}$ for all $i = 0, \ldots, r-1$ and $j = 0, \ldots, s-1$. Thus the two normal series are equivalent and since each is a refinement of one of the original normal series for M, the theorem is proved.

THEOREM 8-10. (Jordan-Hölder Theorem) Any two Jordan-Hölder series for a left Λ-module M are equivalent.

Proof. By Theorem 8-9, the two series have equivalent refinements. Each refinement has the same nonzero factors as the Jordan-Hölder series from which it is obtained. Since the zero factors in the two refinements must correspond, so must the nonzero factors. Hence the nonzero factors, which are in fact all the factors, from the two original series correspond. Thus they are equivalent.

THEOREM 8-11. Let N be a left Λ-submodule of a left Λ-module M. Then

(1) M is left Noetherian if and only if N and M/N are left Noetherian.

(2) M is left Artin if and only if N and M/N are left Artin.

Proof. (1) is just Theorem 6-2 with $M' = N$ and $M'' = M/N$. (2) can be proved by an identical argument, simply reversing all the inclusions.

THEOREM 8-12. Let M be a left Λ-module. The following statements are equivalent:

(1) There exists a Jordan-Hölder series for M.

(2) M is left Artin and left Noetherian.

Proof. (1) implies (2). Suppose M has a Jordan-Hölder series of length 0. Then $M = 0$ and so (2) holds. Assume then that the result holds for modules having Jordan-Hölder series of length $\leq n$. Let $M = M_0 \supset \cdots \supset M_n \supset M_{n+1} = 0$

be a Jordan-Hölder series for M. Then $M/M_n = M_0/M_n \supset \cdots \supset M_{n-1}/M_n \supset M_n/M_n = 0$ is a Jordan-Hölder series for M/M_n of length n. Thus M/M_n is left Artin and left Noetherian. On the other hand $M_n = M_n/M_{n+1}$ is simple and so is left Artin and left Noetherian. By Theorem 8-11, (2) holds for M.

(2) implies (1). Suppose $0 \neq M$ is left Artin and left Noetherian. Let $\mathfrak{S}$ be the collection of all nonzero submodules of M. Since M is left Artin, $\mathfrak{S}$ has a minimal element M_0 which of necessity is a simple submodule of M. Now let $\mathfrak{J}$ be the collection of all nonzero submodules of M which possess Jordan-Hölder series. Then $M_0 \in \mathfrak{J}$ so $\mathfrak{J}$ is nonempty and since M is Noetherian, $\mathfrak{J}$ possesses a maximal element M'. Suppose $M' \neq M$, that is, $M/M' \neq 0$. Since M/M' is left Artin, it too contains simple submodules, that is, there exists a left Λ-module M'' in M such that $M' \subset M'' \subset M$ and M''/M' is a simple left Λ-module. But then M'' must have a Jordan-Hölder series and so is in $\mathfrak{J}$ contradicting the maximality of M'. Thus M' = M has a Jordan-Hölder series.

THEOREM 8-13. (Hopkins' Theorem) Let Λ be a left Artin ring. Then Λ is left Noetherian.

Proof. Let $R = R(\Lambda)$. Then by Theorem 8-5 for some $n \geq 0$, $R^n = 0$. Consider the normal series $\Lambda \supseteq R \supseteq R^2 \supseteq \cdots \supseteq R^n = 0$. We will refine this series into a Jordan-Hölder series for Λ, thereby showing that Λ is left Noetherian.

First consider Λ/R which is again a left Artin ring. Furthermore $R(\Lambda/R) = 0$ so by Theorem 8-7, $\Lambda/R = I_1/R \oplus \cdots \oplus I_n/R$ where $I_1/R, \ldots, I_n/R$ are simple left ideals of Λ/R. Then

$$\Lambda = I_1 + \cdots + I_n \supset I_2 + \cdots + I_n \supset \cdots \supset I_n \supset R$$

is the initial segment of a normal (in fact, Jordan-Hölder) series for Λ. For by Theorem 5-5, $I_j + \cdots + I_n / I_{j+1} + \cdots + I_n \approx I_j / I_j \cap (I_{j+1} + \cdots + I_n) = I_j/R$ which is a simple Λ-module.

It clearly suffices to find for each $i = 1, \ldots, n-1$, a sequence of left ideals of Λ such that $R^i = B_{i,0} \supset \cdots \supset B_{i,m_i} = R^{i+1}$ where each $B_{i,j}/B_{i,j+1}$, $j = 0, \ldots, m_i - 1$ is a simple Λ-module. Fitting all these pieces together, we would then have a Jordan-Hölder series for Λ, completing the proof. For simplicity of notation we will perform this construction for $i = 1$, that is, we will find left ideals $B_0, \ldots, B_m$ of Λ such that $R = B_0 \supset \cdots \supset B_m = R^2$

with B_j/B_{j+1} a simple left Λ-module for $j = 0, \ldots, m-1$.

Now R/R^2 is a left Λ/R-module where the left Λ/R-submodules are of the form A/R^2 where A is a left ideal of Λ with $R^2 \subseteq A \subseteq R$. Since Λ is a left Artin ring, R/R^2 is a left Artin Λ/R-module. Hence by Theorem 8-8(3) $R/R^2 = \bigoplus_{j=0}^{m} A_j/R^2$ where each A_j/R^2 is a simple left Λ/R-submodule of R/R^2. We can therefore construct the next segment of the Jordan-Hölder series, namely,

$$R = A_0 + \cdots + A_m \supset A_1 + \cdots + A_m \supset \cdots \supset A_m \supset R^2.$$

Again Theorem 5-5 can be used to check that the factors are indeed simple Λ-modules.

EXERCISES

8-1. An element $x \in \Lambda$ is called an <u>idempotent</u> if $x^2 = x$. Prove that $R(\Lambda)$ contains no nonzero idempotents.

8-2. Let Λ be the ring of continuous real-valued functions on $[0, 1]$. Prove that $R(\Lambda) = 0$. (See Exercise 2-3.)

8-3. Let Λ be the ring of all 3 x 3 upper triangular matrices with entries in a field k under the usual addition and multiplication of matrices. Compute $R(\Lambda)$ and show that $\Lambda/R(\Lambda)$ is a commutative ring.

8-4. Prove that a finite sum of nilpotent left ideals in a ring Λ is nilpotent. Show by example that a sum of nilpotent elements need not, however, be nilpotent.

8-5. Prove that in a commutative ring R, the sum of a unit and a nilpotent element is a unit.

8-6. Let Λ be a left Artin ring. Prove that every nil left ideal is nilpotent.

8-7. Let $R = k[X]/(f(X))$ where $f(X) \neq 0$. Prove that R is an Artin ring. (Hint: Show that R is a finite dimensional vector space over k.)

8-8. Suppose $\lambda \in \Lambda$ is not nilpotent but $x = \lambda^2 - \lambda$ is nilpotent. Set $\lambda_1 = \lambda + x - 2\lambda x$. Prove

(a) x, λ, and λ_1 commute with one another.

(b) λ_1 is not nilpotent.

(c) $x_1 = \lambda_1^2 - \lambda_1$ is nilpotent.

(d) If $x^n = 0$, then $x_1^{n-1} = 0$.

(e) This process can be iterated (that is, set $\lambda_2 = \lambda_1 + x_1 - 2\lambda_1 x_1$ and proceed as before) to produce an idempotent.

8-9. Suppose A is a non-nilpotent left ideal in Λ such that every proper subideal of A is nilpotent. Suppose further that there exist $\lambda, \mu \in A$ such that $A\mu \neq 0$, $\lambda\mu = \mu$. Prove that λ is not nilpotent but $\lambda^2 - \lambda$ is nilpotent.

8-10. Suppose that Λ is left Artin. Using Exercises 8-8 and 8-9, prove that every non-nilpotent left ideal of Λ contains an idempotent.

CHAPTER 9

SEMISIMPLE RINGS

In this chapter we will study a special class of left Artin rings, namely those whose radical is zero. As we will see, this is equivalent to studying those rings Λ such that every left Λ-module is projective.

DEFINITION 9-1. If M and N are left Λ-modules, we denote by $\text{Hom}_\Lambda(M, N)$ the set of all Λ-homomorphisms from M into N. Under the operations $(f+g)(x) = f(x)+g(x)$ and $(\lambda f)(x) = \lambda f(x)$ for all $x \in M$, $\lambda \in \Lambda$, $\text{Hom}_\Lambda(M, N)$ is endowed with the structure of a left Λ-module. In the case where $M = N$, $\text{Hom}_\Lambda(M, M)$ becomes a ring with unit element under the product $(f \circ g)(x) = f(g(x))$ for all $x \in M$.

THEOREM 9-1. (Schur's lemma) Let E be a simple left Λ-module. Then $\text{Hom}_\Lambda(E, E)$ is a division ring.

Proof. Let $0 \neq f \in \mathrm{Hom}_\Lambda(E, E)$. Then $0 \neq f(E) \subseteq E$ implies $f(E) = E$ so f is surjective. Also $0 \subseteq \ker f \neq E$ implies $\ker f = 0$ so f is injective. Thus f is a Λ-isomorphism and there exists $g \in \mathrm{Hom}_\Lambda(E, E)$ such that $f \circ g = g \circ f = I_E$. Then $g = f^{-1}$.

THEOREM 9-2. Suppose M is a left Λ-module and $\bigoplus_{i=1}^{n} M_i = M = \bigoplus_{j=1}^{m} M'_j$ are two direct sum decompositions of M with M_i, M'_j all simple left Λ-submodules of M. Then $n = m$ and there is a bijection between the M_i and the M'_j in which corresponding modules are isomorphic.

Proof. Suppose with no loss of generality that $n \leq m$. If $n = 1$, M is simple and so $m = 1$. Suppose now that the result holds for $1, 2, \ldots, n-1$. Denote by π'_j the projection from M onto M'_j for each $j = 1, \ldots, m$. Then for some k, $1 \leq k \leq m$, $\pi'_k(M_1) \neq 0$. Otherwise we would have $M_1 = 0$. Since M_1 and M'_k are both simple Λ-modules, it follows that $M_1 \approx M'_k$ and that $(\ker \pi'_k) \cap M_1 = 0$.

It suffices then to show that $N = \bigoplus_{i=2}^{n} M_i \approx \bigoplus_{\substack{j=1 \\ j \neq k}}^{m} M'_j = N'$, for then the induction hypothesis could be applied to complete the proof. But $N' = \ker \pi'_k$ so $N' \cap M_1 = 0$. Thus the sum of M_1 and N' is direct and, since $M_1 \neq 0$, this sum properly contains N'. That is, $N' \subset M_1 \oplus N'$. But

N' is certainly a maximal submodule of M since $M/N' \approx M'_k$ which is simple. Thus $M_1 \oplus N' = M$ from which it follows that $N' \approx M/M_1$. On the other hand, it is immediate that $N \approx M/M_1$ so $N \approx N'$.

THEOREM 9-3. Suppose $\Lambda = \bigoplus_{i=1}^{n} I_i$ is a finite direct sum of simple left ideals and A is a left ideal of Λ. Then A is a finite direct sum of simple left ideals.

Proof. By Theorem 8-8(1), we know $\Lambda = A \oplus B$ and by Theorem 8-8(2), $A = \bigoplus_{j \in J} A_j$ and $B = \bigoplus_{k \in K} B_k$ where A_j and B_k are simple left ideals. Then by Theorem 9-2, $|J| + |K| = n$ so in particular J is a finite set.

THEOREM 9-4. Suppose Λ is a finite direct sum of simple left ideals. Then Λ is left Artin.

Proof. Let $\{A_j \mid j \in J\}$ be a nonempty collection of left ideals of Λ. By Theorems 9-3 and 9-2, each A_j is a finite direct sum of simple left ideals, the number of summands being an invariant of A_j. Choose an ideal A from the collection which has the fewest number of summands when written as a direct sum of simple left ideals. If $B \subset A$ is any left ideal, then by Theorem 8-8(1), $A = B \oplus C$ where $C \neq 0$. Expressing B and C respectively as direct sums of simple left ideals yields a similar expression for A. Thus the number of summands required for B is less than the number for A. Hence $B \notin \{A_j \mid j \in J\}$ and A is a minimal element of the collection. Therefore Λ is left Artin.

DEFINITION 9-2. A ring Λ is called left simple if Λ is left Artin and the only two-sided ideals of Λ are 0 and Λ. Similarly we have the notion of a right simple ring. A ring is called simple if it is both left simple and right simple.

THEOREM 9-5. Let Λ be a division ring and $M_n(\Delta)$ the set of all n x n matrices with entries in Δ. Then $M_n(\Delta)$ is a simple ring.

Proof. We begin by showing that $M_n(\Delta)$ has no two-sided ideals other than 0 and itself. Let A be any nonzero matrix in $M_n(\Delta)$ and let (A) be the two-sided ideal $\{BAC \mid B, C \in M_n(\Delta)\}$. We will show that $(A) = M_n(\Delta)$.

Let $E_{r,s}$ be the matrix with entry 1 in position (r, s) and 0 everywhere else. Suppose $d \neq 0$ is the (i, j) entry of A. Then $X = \Sigma_{k=1}^{m} E_{ki}AE_{jk}$ is in (A) and has d in each position along the diagonal and 0 elsewhere. But the matrix with d^{-1} down the diagonal and 0 elsewhere is the inverse of X. Hence I, the identity matrix, is also in (A), that is,

$(A) = M_n(\Delta)$.

In order to prove that $M_n(\Delta)$ is left (right) Artin, it suffices by Theorem 9-4 to show $M_n(\Delta)$ is a finite direct sum of simple left (right) ideals. Set $M_n(\Delta) = \Lambda$ and $I_j = \Lambda E_{jj}$ for $j = 1, \ldots, n$. I_j is the left ideal of matrices which are zero off the jth column. Clearly $\Lambda = \bigoplus_{j=1}^{n} I_j$ and it remains only to show each I_j is a simple left ideal. Suppose $0 \neq A = (a_{ik}) \in I_j$. Then for some i, $a_{ij} \neq 0$. Let B be the matrix with a_{ij}^{-1} in position (j, i) and 0 elsewhere. Then $BA = E_{jj}$. Thus $I_j = \Lambda E_{jj} = \Lambda BA \subseteq \Lambda A \subseteq I_j$ so $\Lambda A = I_j$. Therefore I_j is a simple left ideal and $M_n(\Delta)$ is a left simple ring. On the other hand, by taking $I_j = E_{jj}\Lambda$, the right ideal of matrices which are zero off the jth row, we get that $M_n(\Delta)$ is right Artin, hence right simple, hence simple.

Thus we have exhibited a whole class of simple rings. One is justified in asking whether there exist any other simple rings and whether there are any rings which are left simple but not right simple or conversely. The answer to all of these questions is no and it is toward these results that we now head.

DEFINITION 9-3. A set of elements $e_1, \ldots, e_n$ in a ring Λ

is called a set of <u>orthogonal idempotents</u> if $e_i e_j = 0$ if $i \neq j$ and $e_i^2 = e_i$. Each e_i is called, naturally, an idempotent.

THEOREM 9-6. The following conditions on a ring Λ are equivalent:

(1) Every short exact sequence of left Λ-modules splits.

(2) Every left Λ-module is Λ-projective.

(3) Every nonzero left Λ-module is the direct sum of simple left Λ-submodules.

(4) Λ is a finite direct sum of simple left ideals. $\Lambda = \oplus_{i=1}^{n} I_i$, where $I_i = \Lambda e_i$ and $\{e_i, \ldots, e_n\}$ is a set of orthogonal idempotents whose sum is 1.

(5) Λ is left Artin and $R(\Lambda) = 0$.

(6) Λ is left Artin and has no nonzero nilpotent left ideals.

(7) Λ is left Artin and $I^2 \neq 0$ for all simple left ideals I.

(8) Λ is left Artin and I is a direct summand of Λ for every simple left ideal I.

(9) Λ is a finite direct sum of rings, $\Lambda = \oplus_{j=1}^{r} \Lambda_j$, where each Λ_j is a two-sided ideal of Λ and is ring isomorphic to $M_{n_j}(\Delta_j)$, the ring of all $n_j \times n_j$ matrices over a division ring Δ_j.

(10) Λ is a finite direct sum of simple rings.

(11) - (18) Replace "left" by "right" in (1) - (8).

Note. Since conditions (9) and (10) make no mention of left or right, it is sufficient by symmetry to prove the equivalence of (1) - (10). A schematic outline of the proof would look like

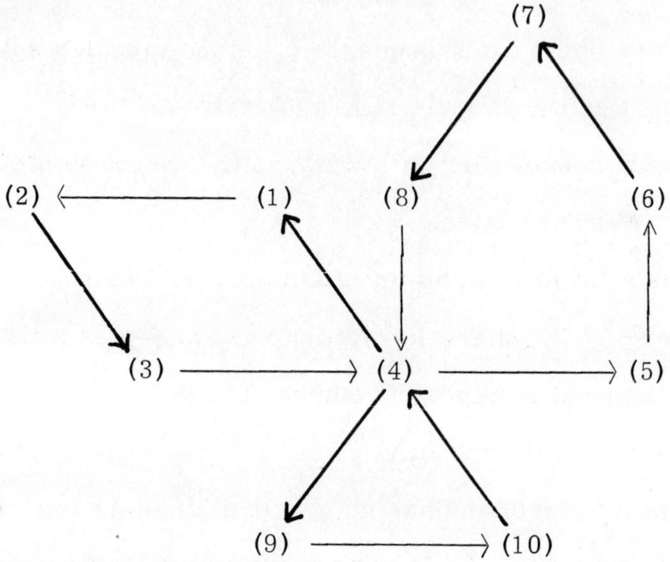

Proof. (4) implies (1). Let $0 \to A \overset{f}{\to} B \overset{g}{\to} C \to 0$ be a short exact sequence of left Λ-modules. By Theorem 8-8(1), we know that f(A) is a direct summand of B, say $B = f(A) \oplus X$. Then $X \approx B/f(A) \approx C$ and $f(A) \approx A$ gives $B \approx A \oplus C$. So the sequence splits.

(1) implies (2). Suppose P is a left Λ-module and we have a Λ-epimorphism $f : B \to C$ and a Λ-homomorphism $g : P \to C$. Then the sequence $0 \to \ker f \to B \overset{f}{\to} C \to 0$ splits so there exists $h : C \to B$ such that $f \circ h = I_C$. Then

$h \circ g : P \to B$ and $f \circ (h \circ g) = g$ so P is Λ-projective.

(2) implies (3). If N is a left Λ-submodule of M, then since M/N is Λ-projective, the sequence $0 \to N \to M \to M/N \to 0$ splits and N is a direct summand of M. The proof of (3) is now identical with that of Theorem 8-8(2).

(3) implies (4). By (3), $\Lambda = \bigoplus_{\alpha \in A} I_\alpha$ where I_α is a simple left ideal of Λ. In particular, $1 = e_1 + \cdots + e_n$ where $e_i \in I_{\alpha_i}$. So for any $\lambda \in \Lambda$, $\lambda \cdot 1 = \lambda e_1 + \cdots + \lambda e_n$. Therefore $\Lambda \subseteq \bigoplus_{i=1}^n I_{\alpha_i} \subseteq \bigoplus_{\alpha \in A} I_\alpha = \Lambda$ so $\Lambda = \bigoplus_{i=1}^n I_i$ where $I_i = I_{\alpha_i}$. If $x_i \in I_i$, then $x_i = x_i 1 = x_i e_i + \cdots + x_i e_n$. Also $x_i = 0 + \cdots + 0 + x_i + 0 + \cdots + 0$. Thus $x_i e_i = x_i$, $x_i e_j = 0$ for $i \neq j$. Hence $I_i = \Lambda e_i$. Also setting $x_i = e_i$, we get that $\{e_1, \ldots, e_n\}$ is a set of orthogonal idempotents.

(4) implies (5). By Theorem 9-4, we have that Λ is left Artin. Also $\mathfrak{m}_i = \bigoplus_{\substack{j=1 \\ j \neq i}}^n I_j$ is a maximal left ideal of Λ since $\Lambda/\mathfrak{m}_i = I_i$ is a simple left Λ-module. Thus $R(\Lambda) \subseteq \bigcap_{i=1}^n \mathfrak{m}_i = 0$ so $R(\Lambda) = 0$.

(5) implies (6). This follows from the Corollary to Theorem 8-3.

(6) implies (7). If $I^2 = 0$, then I is nilpotent so $I = 0$. Thus I is not simple.

(7) implies (8). This is simply Theorem 8-6.

(8) implies (4). If one looks at the proof of Theorem 8-7, the hypothesis that $R(\Lambda) = 0$ was used only to prove that a certain simple left ideal I was a direct summand of Λ which is a hypothesis here. Thus by Theorem 8-7, we have $\Lambda = \oplus_{i=1}^{n} I_i$ for simple left ideals I_i. Expressing $1 = e_1 + \cdots + e_n$ with $e_i \in I_i$, we get immediately that $I_i = \Lambda e_i$ and $\{e_1, \ldots, e_n\}$ is a set of orthogonal idempotents.

(9) implies (10). This is immediate from Theorem 9-5.

(10) implies (4). Suppose $\Lambda = \oplus_{j=1}^{r} \Lambda_j$ where each Λ_j is a simple ring. Then Λ_j is left Artin and since $R(\Lambda_j)$ is a two-sided ideal of Λ_j either $R(\Lambda_j) = \Lambda_j$ or $R(\Lambda_j) = 0$. The first possibility is ruled out by the existence of a maximal left ideal in Λ_j, an easy consequence of Zorn's lemma. Thus Λ_j satisfies (5) or equivalently (4). Suppose $\Lambda_j = \oplus_{i=1}^{n} I_{ij}$ where $I_{ij} = \Lambda_j e_{ij}$ is a simple left ideal of Λ_j, $\Sigma_{i=1}^{n_j} e_{ij} = 1_j$, and $\{e_{ij}\}$, $i = 1, \ldots, n_j$, is a set of orthogonal idempotents. Then $\Lambda = \oplus_{i,j} I_{ij}$ where $I_{ij} = \Lambda e_{ij}$ is a simple left ideal of Λ, $\{e_{ij}\}$, $i = 1, \ldots, n_j$, $j = 1, \ldots, r$, is a set of orthogonal idempotents, and $\Sigma_{i,j} e_{ij} = \Sigma_{j=1}^{r} 1_j = 1$. Thus Λ satisfies (4).

(4) implies (9). We have saved the most intricate and the only really new part of the proof to last. Suppose then that $\Lambda = \oplus_{i=1}^{m} I_i$ where $I_i = \Lambda e_i$ is a simple left ideal

and $\{e_1, \ldots, e_m\}$ is a set of orthogonal idempotents whose sum is 1. We regroup these simple left ideals as follows: Let Λ_1 be the direct sum of all I_i which are isomorphic to I_1 as left Λ-modules. If I_{i_2} is the first ideal not isomorphic to I_1, let Λ_2 be the direct sum of all those I_i which are isomorphic to I_{i_2} as left Λ-modules. Continue in this way to get $\Lambda = \bigoplus_{j=1}^{r} \Lambda_j$ where each Λ_j is a direct sum of isomorphic simple left ideals and if $i \neq j$, no summand of Λ_i is isomorphic to a summand of Λ_j.

Our first claim is that each Λ_j is a two-sided ideal of Λ. We know that Λ_j is a left ideal since it is a direct sum of left ideals. We must show $\Lambda_j\Lambda = \Lambda_j(\Lambda_1 \oplus \cdots \oplus \Lambda_r) \subseteq \Lambda_j$. It is clearly sufficient to show that $\Lambda_j\Lambda_k = 0$ for $j \neq k$. And to show this it is sufficient to show that $IJ = 0$ for any two nonisomorphic simple left ideals I and J of Λ. Suppose then that $IJ \neq 0$. Then there exists $y \in J$ such that $Iy \neq 0$. Define a Λ-homomorphism $\tilde{y} : I \to J$ by $\tilde{y}(x) = xy$ for all $x \in I$. Then $\tilde{y} \neq 0$ and so must be an isomorphism between I and J. Thus the claim is verified.

We wish to show now that each Λ_j is isomorphic as a ring to $M_{n_j}(\Delta_j)$ for some appropriately chosen division ring Δ_j. In order to simplify the notation, suppose that $\Lambda = \bigoplus_{i=1}^{n} I_i$ where the I_i are simple left ideals and $I_1 \approx I_2 \approx \cdots \approx I_n$ as left Λ-modules. Thus we are looking at one of the summands Λ_j. Set $\Delta = \mathrm{Hom}_\Lambda(I_1, I_1)$ which by Schur's lemma is a division ring. We will prove $\Lambda \approx M_n(\Delta)$.

We begin by investigating the structure of $\mathrm{Hom}_\Lambda(I_i, I_j)$, asking whether there is some easy way of viewing any Λ-homomorphism $f : I_i \to I_j$. Now $f(e_i) = \lambda e_j$ for some $\lambda \in \Lambda$. Thus $f(e_i) = f(e_i^2) = e_i f(e_i) = e_i \lambda e_j$. But for any $x \in I_i$, $x = xe_i$, so $f(x) = f(xe_i) = xf(e_i) = xe_i\lambda e_j$. Therefore $f = \widetilde{e_i\lambda e_j}$ in our usual notation where $\widetilde{z}$ denotes the map given by right multiplication by z.

Let us pursue this fact to determine how these maps are added and composed. Suppose f, $g \in \mathrm{Hom}_\Lambda(I_i, I_j)$ where $f = \widetilde{e_i\lambda e_j}$ and $g = \widetilde{e_i\mu e_j}$. Then $(f+g)(x) = f(x)+g(x)$ $= xe_i\lambda e_j + xe_i\mu e_j = xe_i(\lambda+\mu)e_j$ so $f+g = \widetilde{e_i(\lambda+\mu)e_j}$. Now suppose $f \in \mathrm{Hom}_\Lambda(I_i, I_j)$ and $g \in \mathrm{Hom}_\Lambda(I_j, I_k)$ where $f = \widetilde{e_i\lambda e_j}$ and $g = \widetilde{e_j\mu e_k}$. Then $g \circ f \in \mathrm{Hom}_\Lambda(I_i, I_k)$ and $(g \circ f)(x) = g(f(x)) = g(xe_i\lambda e_j) = xe_i\lambda e_j e_j \mu e_k = xe_i\lambda e_j\mu e_k$. Thus $g \circ f = \widetilde{e_i\lambda e_j\mu e_k}$. Note that the "multipliers" appear in the reverse order to the usual composition of functions.

For convenience let us drop the $\sim$ notation, that is, $e_i\lambda e_j$ is now to be thought of as an element of $\mathrm{Hom}_\Lambda(I_i, I_j)$ given by right multiplication by $e_i\lambda e_j$. Similarly, compose functions from left to right rather than in our usual way to agree with the way these homomorphisms are multiplied.

Let g_i be a nonzero element of $\mathrm{Hom}_\Lambda(I_1, I_i)$ and

$g_i^{-1} \in \mathrm{Hom}_\Lambda(I_i, I_1)$ be its inverse. Choose such homomorphisms for $i = 1, \ldots, n$. Then $g_i e_i \lambda e_j g_j^{-1} \in \mathrm{Hom}_\Lambda(I_1, I_1) = \Delta$ for each $\lambda \in \Lambda$. We thus define a map $\rho : \Lambda \to M_n(\Delta)$ by $\rho(\lambda) = | g_i e_i \lambda e_j g_j^{-1} |$, the matrix whose (i, j) entry is $g_i e_i \lambda e_j g_j^{-1}$. We will prove that ρ is a ring isomorphism.

(1) ρ is a homomorphism.

$$\rho(\lambda_1 + \lambda_2) = |g_i e_i(\lambda_1 + \lambda_2)e_j g_j^{-1}|$$

$$= |g_i e_i \lambda_1 e_j g_j^{-1}| + |g_i e_i \lambda_2 e_j g_j^{-1}|$$

$$= \rho(\lambda_1) + \rho(\lambda_2);$$

$$\rho(\lambda_1 \lambda_2) = |g_i e_i(\lambda_1 \lambda_2)e_j g_j^{-1}|$$

$$= |g_i e_i \lambda_1 (\Sigma_{k=1}^n e_k)\lambda_2 e_j g_j^{-1}|$$

$$= |\Sigma_{k=1}^n g_i e_i \lambda_1 e_k \lambda_2 e_j g_j^{-1}|$$

$$= |\Sigma_{k=1}^n (g_i e_i \lambda_1 e_k g_k^{-1})(g_k e_k \lambda_2 e_j g_j^{-1})|$$

$$= \rho(\lambda_1)\rho(\lambda_2).$$

(2) ρ is injective. For suppose $\lambda \neq 0$. Then for some values of i and j, $e_i \lambda e_j \neq 0$. Otherwise $0 = \Sigma_{i,j=1}^n e_i \lambda e_j = \Sigma_{j=1}^n (\Sigma_{i=1}^n e_i \lambda) e_j = \Sigma_{j=1}^n \lambda e_j = \lambda$, contradiction. Thus $e_i \lambda e_j$ is a Λ-isomorphism from I_i to I_j and so $g_i e_i \lambda e_j g_j^{-1}$ is the composition of three isomorphisms, hence

is an isomorphism from I_1 to I_1, that is, is a nonzero element of Δ. Thus $\rho(\lambda) \neq 0$.

(3) ρ is surjective. It is clearly enough to show that for each element $\delta \in \Delta$ and pair of indices (i, j) the matrix with δ in position (i, j) and 0 elsewhere is $\rho(\lambda)$ for some $\lambda \in \Lambda$. Now $g_i^{-1}\delta g_j \in \mathrm{Hom}_\Lambda(I_i, I_j)$ so $g_i^{-1}\delta g_j = e_i\lambda e_j$ for some $\lambda \in \Lambda$. Thus $\delta = g_i e_i \lambda e_j g_j^{-1}$. Therefore the (k, ℓ) entry of $\rho(e_i\lambda e_j)$ is $g_k e_k e_i \lambda e_j e_\ell g_\ell^{-1}$ which is equal to δ if $k = i$ and $\ell = j$ and 0 otherwise.

DEFINITION 9-4. A ring satisfying any (hence all) of the above conditions is called <u>semisimple</u>. Note that we need not distinguish between left and right semisimple since a ring having one property necessarily has both.

COROLLARY 1. (1) Every simple ring is semisimple. (2) A ring is left simple if and only if it is right simple.

COROLLARY 2. (Wedderburn's Theorem) Every simple ring Λ is isomorphic to $M_n(\Delta)$, the ring of all n x n matrices over a division ring Δ.

<u>Proof</u>. By Corollary 1(1), Λ is semisimple so by condition (9) of Theorem 9-6, $\Lambda = \bigoplus_{j=1}^{r} \Lambda_j$ where Λ_j is a

two-sided ideal of Λ isomorphic to $M_{n_j}(\Delta_j)$. But the only two-sided ideals of Λ are 0 and Λ. Thus r = 1 and $\Lambda \approx M_n(\Delta) = M_{n_1}(\Delta_1)$.

Finally we have the question of the uniqueness of the decompositions of conditions (9) and (10) of the theorem. We conclude with the following uniqueness theorem.

THEOREM 9-7. Suppose $\bigoplus_{i=1}^{n} \Lambda_i = \Lambda = \bigoplus_{j=1}^{m} \Lambda_j'$, are two ring direct sum decompositions of Λ with Λ_i and Λ_j' simple rings. Then n = m and each Λ_i equals some Λ_j'.

<u>Proof.</u> Since Λ has a unit element, we have $\Lambda_i = \Lambda_i\Lambda = \Sigma_{j=1}^{m} \Lambda_i\Lambda_j'$. Exactly one of these summands is nonzero. For clearly they cannot all be 0 since $\Lambda_i \neq 0$. On the other hand, since each $\Lambda_i\Lambda_j'$ is a two-sided ideal of Λ_i, if it is not 0, it must be Λ_i. But if $\Lambda_i\Lambda_j' = \Lambda_i = \Lambda_i\Lambda_k'$, then $\Lambda_j' \cap \Lambda_k' \supseteq \Lambda_i \neq 0$, contradicting the direct sum property. Thus for exactly one value of j, $\Lambda_i = \Lambda_i\Lambda_j'$. But $\Lambda_i\Lambda_j'$ is also a two-sided ideal of Λ_j' and since it is nonzero must be Λ_j'. Therefore $\Lambda_i = \Lambda_j'$.

EXERCISES

9-1. Prove that every left Artin ring with no nontrivial

zero divisors is a field.

9-2. Prove that if e is an idempotent in a ring Λ, then Λe is a projective Λ-module.

9-3. Let Λ be the ring of quaternions over $\mathbb{Z}_3$, that is, $\Lambda = \{x_0 + x_1 i + x_2 j + x_3 k \mid x_0, x_1, x_2, x_3 \in \mathbb{Z}_3\}$ where the operations in Λ are described in Chapter 1, Example 7. Prove that Λ is a simple ring. (Hint: Exhibit an isomorphism between Λ and $M_2(\mathbb{Z}_3)$.)

9-4. Let Λ be a semisimple ring. Prove that every simple Λ-module is isomorphic to a simple left ideal of Λ.

9-5. Let Λ be a simple ring. Prove that all simple Λ-modules are isomorphic.

9-6. Let Λ be a semisimple ring and A and B simple left Λ-modules. Prove that $A \approx B$ if and only if $(0:A) = (0:B)$.

9-7. Prove that any two-sided ideal in a semisimple ring is itself a semisimple ring.

9-8. Show that a commutative semisimple ring is a finite

direct sum of fields.

9-9. Let $C(\Lambda)$ denote the center of the ring Λ (see Exercise 1-5). Prove that if $R(\Lambda) = 0$, then $R(C(\Lambda)) = 0$.

9-10. Suppose Λ is a semisimple ring. Describe $C(\Lambda)$ in terms of matrices.

BIBLIOGRAPHY

1. M. Auslander, Rings, Modules, and Homology, Brandeis University Lecture Notes, Waltham, Mass., 1960.
2. N. Jacobson, Lectures in Abstract Algebra, Van Nostrand, Princeton, N. J., Vol. 1, 1951, Vol. 2, 1953.
3. N. Jacobson, Structure of Rings, American Mathematical Society Colloquium, Vol. 37, Providence, R. I., 1956.
4. J. P. Jans, Rings and Homology, Holt, Rinehart and Winston, New York, 1964.
5. J. Lambek, Lectures on Rings and Modules, Blaisdell, Waltham, Mass., 1966.
6. S. Lang, Algebra, Addison-Wesley, Reading, Mass., 1965.
7. N. H. McCoy, The Theory of Rings, Macmillan, New York, 1964.
8. D. G. Northcott, Ideal Theory, Cambridge University Press, Cambridge, England, 1953.
9. L. van der Waerden, Modern Algebra, Ungar, New York, 1949.
10. O. Zariski and P. Samuel, Commutative Algebra, Vol. 1, Van Nostrand, Princeton, N. J., 1958.

INDEX

Abelian group, 5
Addition of ideals, 13
Adic topology, 90
Annihilator of a module, 74
Artin module, 110
Artin ring, 110
Artin-Rees theorem, 89
Ascending chain condition, 39
Assassinator of a module, 76
Associated ideals, 72
Associated primes of a module, 76

Basis of a module, 55
Basis of an ideal, 14
Bijective, 4

Center of a ring, 10
Chain, 3
Chinese remainder theorem, 23
Cokernel, 65
Comaximal ideals, 23
Commutative ring, 6
Commutative ring with unit, 6
Congruent modulo, 17
Content of a polynomial, 49
Contraction of an ideal, 31
Cross-section, 58

Dedekind domain, 102
Descending chain condition, 110
Direct sum of modules, 53
Direct sum of rings, 22
Direct summand, 53
Discrete valuation, 104
Discrete valuation ring, 104
Division ring, 6
Domain, 20

Eisenstein's criterion, 50
Embedded prime ideal, 87
Epimorphism, 7
Equivalence class, 1
Equivalence relation, 1
Equivalent normal series, 116
Euclidean domain, 38
Exact sequence, 56
Extension of an ideal, 31

Factor module, 53
Factor theorem, 50
Factors of a normal series, 116

Field, 6
Finitely generated ideal, 14
Finitely generated module, 54
First isomorphism theorem, 18
Five lemma, 64
Fractionary ideal, 95
Free module, 56

Gauss lemma, 49
Gaussian integers, 10
Generators of a module, 55
Generators of an ideal, 14
Greatest common divisor, 49
Group, 4

Hilbert basis theorem, 71
Hom, 124
Homomorphism of groups, 5
Homomorphism of modules, 54
Homomorphism of rings, 6
Hopkins' theorem, 119

Ideal, 12
Idempotent, 121
Identity element, 4
Image, 16
Inductive set, 3
Injective, 4
Integral domain, 20
Integral element, 104
Integral ideal, 96
Integrally closed domain, 104
Intersection of ideals, 13
Inverse element, 5
Invertible ideal, 96
Irreducible element, 41
Isolated prime ideal, 87
Isomorphism, 7

Jordan-Hölder series, 116
Jordan-Hölder theorem, 117

Kernel, 16
Krull intersection theorem, 90

Lasker-Noether decomposition, 83
Length of a normal series, 116
Local ring, 34
Localization, 30

Matrix ring, 8
Maximal element, 3
Maximal ideal, 16
Maximum condition, 40
Minimal ideal, 111
Minimum condition, 110
Module, 52
Module of quotients, 61
Monomorphism, 7
Multiplication of ideals, 13
Multiplicative inverse, 6
Multiplicative set, 28

Nakayama's lemma, 108
Nil ideal, 109
Nil radical of an ideal, 81
Nilpotent ideal, 109
Noetherian module, 69
Noetherian ring, 69
Nontrivial zero divisor, 20
Normal series, 116

Orthogonal idempotents, 129

Partial ordering, 2
Partially ordered set, 3
Polynomial ring, 7
Primary decomposition of zero, 79
Primary ideal, 76
Primary module, 76
Prime element, 41
Prime ideal, 15
Prime module, 75
Primitive polynomial, 49
Principal ideal, 14
Principal ideal ring, 14
Projective module, 60

Proper ideal, 13

Quaternions, 8
Quotient field, 30
Quotient of ideals, 13

Radical of a ring, 108
Refinement of a normal series, 116
Regular ring, 51
Residue ring, 18
Retraction, 58
Ring, 5
Ring of quotients, 29
Ring with unit, 6

Schreier refinement theorem, 116
Schur's lemma, 124
Second isomorphism theorem, 19
Semisimple ring, 136
Short exact sequence, 56
Simple ideal, 111
Simple module, 106
Simple ring, 127
Snake diagram, 65
Split exact sequence, 58
Submodule, 53
Surjective, 4

Total ring of quotients, 30
Totally ordered set, 3

Unique factorization domain, 41
Unit, 33
Unit element, 6
Unitary homomorphism, 7
Upper bound, 3

Wedderburn's theorem, 136

Zero divisor, 20
Zorn's lemma, 3